Risk and the Public Acceptance of New Technologies

Also by Rob Flynn

CONTRACTING FOR HEALTH (*co-editor*)
MARKETS AND NETWORKS (*co-author*)
STRUCTURES OF CONTROL IN HEALTH MANAGEMENT

Also by Paul Bellaby

SICK FROM WORK

Risk and the Public Acceptance of New Technologies

Edited by

Rob Flynn and Paul Bellaby
Institute for Social, Cultural and Policy Research
University of Salford

First published 2007 by
PALGRAVE MACMILLAN
Houndmills, Basingstoke, Hampshire RG21 6XS and
175 Fifth Avenue, New York, N.Y. 10010
Companies and representatives throughout the world

PALGRAVE MACMILLAN is the global academic imprint of the Palgrave Macmillan division of St. Martin's Press, LLC and of Palgrave Macmillan Ltd. Macmillan® is a registered trademark in the United States, United Kingdom and other countries. Palgrave is a registered trademark in the European Union and other countries.

ISBN 13: 978–0–230–51705–9   hardback

This book is printed on paper suitable for recycling and made from fully managed and sustained forest sources. Logging, pulping and manufacturing processes are expected to conform to the environmental regulations of the country of origin.

A catalogue record for this book is available from the British Library.

Library of Congress Cataloging-in-Publication Data

Risk and the public acceptance of new technologies / Edited by Rob Flynn and Paul Bellaby.
     p. cm.
   Includes bibliographical references and index.
     ISBN 978-0-230-51705-9 (alk. paper)
     1. Technological innovations–Social aspects. I. Flynn, Rob. II. Bellaby, Paul.

   T173.8.R57 2007
   303.48'3–dc22   2007022512

10   9   8   7   6   5   4   3   2   1
16   15   14   13   12   11   10   09   08   07

Transferred to Digital Printing 2011

Contents

List of Figures and Tables

Figures

Tables

Notes on Contributors

Dr. Julie Barnett is Senior Research Fellow, Department of Psychology, University of Surrey, UK. Her research interests concern changing patterns of appreciation of risks in publics, organisations and institutions; communicating risk; expert understandings of publics; processes of public engagement and dialogue; and the role of public and stakeholder engagement in evidence-based policy-making. Her recent major publications include: Barnett J. and Breakwell G. (2003) 'The social amplification of risk and the hazard sequence', *Health, Risk and Society*, 5, 3, 301–313; Barnett J., Carr, A. and Clift, R. (2006), 'Going public: risk, trust and public understandings of nanotechnology', in G. Hunt and M. Mehta (eds), *Nanotechnology: Risk, Ethics and Law*; and Timotijevic, L. and Barnett, J. (2006) 'Managing the possible health risks of mobile telecommunications', *Health, Risk and Society*, 8, 2, 143–164.

Professor Paul Bellaby is Professor of Sociology, and Director of the Institute for Social, Cultural and Policy Research, University of Salford, Greater Manchester, UK. He took his first degree and PhD at University of Cambridge, and has held posts at the Universities of Keele and East Anglia. His current research interests lie in the sociology of health, such new technologies as hydrogen energy and the internet, and how risk is conceived and trust is at issue between science, policy makers and publics. Publications include (2007) with Eames, M. (eds) special issue of *Energy Policy* on 'Trust among stakeholders in managing uncertainties on the way to sustainable energy'; (2006) 'Can they carry on working? Later retirement, health, and social inequality in an aging population', *International Journal of Health Services*, 36.1: 1–23; (2003) 'Communication and miscommunication of risk: understanding UK parents' attitudes to combined MMR vaccination', *British Medical Journal*, Vol. 327: 725–728; (1999) *Sick from Work*.

Dr. Hannah Devine-Wright is a Research Fellow in environmental psychology at the University of Manchester, UK working on social aspects of evolving electricity systems and demand-side participation, within the EPSRC-funded 'Supergen' Future Networks project. She holds Master's and PhD degrees from the University of Surrey and is a graduate member of the British Psychological Society. Her multi-methods research addresses social and design issues associated with novel or existing technologies such as 'smart' meters or electricity pylons. She has a special interest in theoretical and applied aspects of visual methods as a means of exploring the symbolic meaning of these technologies. Her publications include: H. Devine-Wright and P. Devine-Wright (2004) 'From demand-side management to demand-side participation: towards an environmental psychology of sustainable electricity system evolution', *Journal of Applied Psychology*, 6, 3–4, 167–177; P. Devine-Wright and H. Devine-Wright (2006), 'Social representations of intermittency and the shaping of public support for wind energy in the UK', *International Journal of Global Energy Issues*, 25, 3–4, 243–256; and H. Devine-Wright and P. Devine-Wright (2006) 'Prospects for smart metering in the UK', in T. Jamash, M. Pollitt and W. Nuttall (eds), *Future Technologies for a Sustainable Electricity System*.

Dr. Patrick Devine-Wright is a Senior Lecturer at the University of Manchester Architecture Research Centre, UK. With a background in environmental and social psychology, his main research interests lie in the social acceptance of sustainable energy technologies, motivation for pro-environmental behaviour and place theory. He is involved in several major research projects funded by the UK Government, including leading the 'Beyond NIMBYism' interdisciplinary research project on public engagement with renewable energy technologies. He is a member of the International Science Panel on Renewable Energy, has sat on the National Advisory Group steering the Community Renewables Initiative since 2001, and is a board member for the journal *Local Environment*. He is a chartered psychologist and graduate member of the British Psychological Society. Publications include: P. Devine-Wright (2005) 'Local aspects of renewable energy development in the UK: public beliefs and policy implications', *Local Environment*, 10, 1, 57–69; P. Devine-Wright (2005) 'Beyond NIMBY-ism: towards an integrated framework for understanding public perceptions of wind energy', *Wind Energy*, 8, 2, 125–139; P. Devine-Wright (2007) 'Energy citizenship: psychological aspects of evolution in sustainable energy technologies', in J. Murphy (ed.), *Framing the Present, Shaping the Future: contemporary governance of sustainable technologies*.

Professor Malcolm Eames is Professor of Innovation and Sustainable Development at the Brunel Business School, Brunel University, UK. He was previously Senior Research Fellow at the Policy Studies Institute, London, where he led the UK Sustainable Hydrogen Consortium's work on hydrogen future scenarios. He was also co-ordinator of the Sustainable Development Research Network. His current research interests span: sustainable development research policy; participatory and deliberative decision-making; socio-economic and technological scenario-building; technology assessment; and energy and environmental futures, and environmental justice. His major recent publications include: M. Eames and J. Skea (2003) 'The development and use of the UK Environmental Futures Scenarios', *Greener Management International*, 37, 53–70; M. Eames *et al.* (2004) 'Xenotransplantation and its alternatives for addressing the "kidney gap": a case-study from the deliberative mapping project', Report from the Policy Studies Institute, London; and M. Eames, W. McDowall, M. Hodson and S. Marvin (2006), 'Negotiating generic and place-specific expectations of a hydrogen economy', *Technology Analysis and Strategic Management*, 18, 361–374.

Dr. Arnout Fischer is a postdoctoral researcher in the Marketing and Consumer Behaviour group at Wageningen University in the Netherlands. His research concerns understanding consumer food practices and choices, using psychological models that interpret human behaviour as the results of heuristics, experience, emotion and automatic behaviour. His research interests also include the influence of risk-benefit perceptions on behaviour and how consequences of behaviour influence future behaviour. Recent publications include: Fischer, A.R.H.; Frewer, L.J.; Nauta, M.J. (2006) 'Toward Improving Food Safety in the Domestic Environment: A Multi-Item Rasch Scale for the Measurement of the Safety Efficacy of Domestic Food-Handling Practices' *Risk Analysis*, 26, 1323–1338; and Fischer, A.R.H.; Jong, A.E.I. de; Jonge, R. de; Frewer, L.J.; Nauta, M.J. (2005) 'Improving Food Safety in the Domestic Environment: The Need for a Transdisciplinary Approach', *Risk Analysis*, 25, 503–517.

Professor Rob Flynn is Professor of Sociology at the University of Salford, UK. His research interests include public perceptions of risk and trust in experts; public engagement in science and technology; governance and the regulation of professionals; and health policy and health services organisation. He was a lead co-investigator in the EPSRC-funded UK Sustainable Hydrogen Energy Consortium, working collaboratively with several contributors to this volume on socio-economic aspects of a hydrogen economy. Together with Paul Bellaby, he has been a member of a Department of Transport-funded research team investigating public engagement in future transport systems based on hydrogen infrastructures. He has previously been chairperson of the editorial board of *Sociology*, and is currently Chairperson of the editorial board of *Sociology of Health and Illness*. His major publications include: R. Flynn (2002) 'Clinical governance and governmentality', *Health, Risk and Society*, 4, 2, 155–173; R. Flynn, P. Bellaby and M. Ricci (2006) 'Risk perception of an emergent technology: the case of Hydrogen energy', *Forum for Qualitative Research*, 7, 1: online at http://www.qualitative-research.net/fqs/fqs-e/inhalt1-06-e.htm; and R. Flynn (2006) 'Risk and health', in G. Mythen and S. Walklate (eds), *Beyond the Risk Society*.

Professor Lynn Frewer is Professor in Food Safety and Consumer Behaviour at Wageningen University in the Netherlands. Her research interests include the application of psychological models in understanding consumer attitudes to emerging food technologies, developing and testing models of risk/benefit perception, and communication, and investigating their subsequent impacts on food choice behaviour. Professor Frewer also investigates the importance of public trust and institutional credibility in food risk governance, and is particularly interested in developing collaborations between the natural and social sciences. Lynn has published extensively on these topics, including for example: S. Miles and L.J. Frewer (2003) 'Public perception of scientific uncertainty in relation to food hazards' *Journal of Risk Research*, 6, 3, 267–283; G. Rowe and L.J. Frewer (2004), 'A typology of public engagement mechanisms', *Science, Technology and Human Values*, 29, 4, 512–290; and van Kleef, E., Frewer, L.J., Chryssochoidis, G., Houghton, J.R., Korzen-Bohr, S., Crystallis, T. (2006), 'Perceptions of food risk management among key stakeholders: Results from a cross European study', *Appetite*, 47, 46–63.

Dr. Mike Hodson is Research Fellow at the Centre for Sustainable Urban and Regional Futures (SURF), University of Salford, UK. His research interests focus broadly on relationships between socio-technologies and contexts of innovation. He was a member of the EPSRC-funded UK Sustainable Hydrogen Energy Consortium (UKSHEC), principally addressing transitions to city and regional hydrogen economies. He was co-organiser of the ESRC Seminar Series, 'Analysing Social Dimensions of Emerging Hydrogen Economies' (2005–6). His recent publications include: Hodson, M. and Marvin, S. (2006), 'Reconnecting the Technology Characterisation of the Hydrogen Economy to Contexts of Consumption', *Energy Policy*, 34, 3006–3016; and Hodson, M. (2007) 'Old Industrial Regions, Technology and Innovation: Tensions of Obduracy and Transformation', *Environment and Planning A*.

Dr. Tom Horlick-Jones is an independent researcher and consultant, currently based at Cardiff School of Social Sciences, Cardiff University, UK. Over a period of some twenty years he has specialised in issues concerned with applied and conceptual aspects of risk, organisations and decision-making processes. His research work is much concerned with the roles of practical reasoning, language and knowledge in these areas. His publications include: T. Horlick-Jones, A. Amendola, and R. Cassale (eds) (1995) *Natural Risk and Civil Protection* (1995); J. Petts, T. Horlick-Jones and G. Murdock (2001) *Social Amplification of Risk: the Media and the Public* and T. Horlick-Jones, J. Walls, G. Rowe, N. Pidgeon, W. Poortinga, W. Murdock and T. O'Riordan (2007) *The GM Debate: Risk, Politics and Public Engagement.*

Professor Alan Irwin was Professor of Science and Technology Policy at the University of Liverpool, UK, and is now Dean of Research, Copenhagen Business School, Denmark. His current research interests are in scientific governance; science-public relations; and Science and Technology Studies (STS). He sits on the public engagement advisory committees of the Wellcome Trust and the BBSRC (Biotechnology and Biological Sciences Research Council, UK). He has published extensively about the sociology of science and technology, and social science and environmental issues. His publications include: A. Irwin (1995) *Citizen Science*; A. Irwin (2001) *Sociology and the Environment*; A. Irwin and M. Michael (2003), *Science, Social Theory and Public Knowledge*; and J. Stilgoe, A. Irwin and K. Jones (2006) *The Received Wisdom: opening up expert advice.*

Professor Simon Marvin is Professor and Director of the Centre for Sustainable Urban and Regional Futures, University of Salford, UK. His main research interests are in understanding the changing relations between cities, regional and infrastructure networks. Recent work has focused on understanding interdisciplinarity in urban research and the role of cities and regions in shaping systemic change in socio-technical networks. With Mike Hodson, he was a co-organiser of the ESRC Seminar Series 'Analysing Social Dimensions of Emerging Hydrogen Economies' (2005–6). Recent books include Guy, S., Marvin, S.J. and Moss, T. (2001), *Urban Infrastructure in Transition: Networks, Buildings and Plans*, and Graham, S.D.N. and Marvin, S.J. (2001), *Splintering Urbanism: Networked Infrastructures, Technological Mobilities and the Urban Condition.*

Will McDowall is a researcher at the Centre for Health and Environmental Research, University of British Columbia, Canada.

Previously he was Research Fellow in the environment group at Policy Studies Unit, London, where he was part of the UK Sustainable Hydrogen Energy Consortium. His undergraduate education was in biology, and his postgraduate training was in biodiversity and conservation. His main research interests are long-term technological change; participatory technology assessment; science communication and the uses of science in policy. Currently he is working on knowledge translation and exchanges between scientists and policy-makers. Recent publications include: P. Agnolucci and W. McDowall 'Technological change in niches: the auxiliary power unit and the hydrogen economy', *Technological Forecasting and Social Change*; and W. McDowall and M. Eames (2006), 'Forecasts, scenarios, visions, backcasts and roadmaps to the hydrogen economy', *Energy Policy*, 34, 1236–1250.

Dr. Alison Mohr is Senior Research Fellow at the Institute for Science and Society, University of Nottingham, UK. Her research interests include participatory governance processes in science and technology policy, and the social construction of new technologies. She was a partner in the European Commission-funded 'NanoDialogue' project which brought together European policy-makers, scientific experts and members of the public to consider the economic, environmental, ethical and social issues in nanotechnologies. Among her publications are: J. Walls, T. Rogers-Hayden, A. Mohr and T. O'Riordan (2005) 'Seeking citizens' views on GM crops: experiences from the UK, Australia and New Zealand', *Environment*, 47, 7, 22–36; and she is co-editor (with T. Rogers-Hayden, D. Guston, N. Pidgeon and B. Wynne) of a forthcoming book *Engaging with nanotechnologies – engaging differently?*

Dr. Susana Mourato is Senior Lecturer in Environmental Economics at Imperial College, London, UK. Her research focuses on the application of economic valuation techniques to the measurement of environmental, social and cultural changes. Recent research includes measuring preferences for hydrogen and bio-fuels transport technologies; valuing the impacts of transport infrastructure on the natural landscape; the benefits of reducing the risk of water shortages and sewage overflows; and estimating the non-market impacts of the 2012 London Olympics. Among her recent publications are: D. Pearce, G. Atkinson and S. Mourato (2006) *Cost-Benefit Analysis and the Environment: recent developments* (Paris: OECD, and Edward Elgar: Cheltenham); and with I. Bateman *et al.* (2002) *Economic Valuation with Stated Preference techniques: a manual* and numerous journal articles on economic valuation.

Dr. Tanya O'Garra is currently postdoctoral Research Fellow at the University of the South Pacific in Fiji, and is also a Visiting Research Associate at Imperial College, London. Her PhD studies investigated public acceptance of, and preferences for hydrogen buses and associated refueling infrastructure. She is also involved in a BP-funded study of public attitudes to hydrogen infrastructure in London; an EPSRC-funded project on hydrogen refuelling infrastructure; and an international project on public acceptance of hydrogen buses ('AcceptH2'). Her main publications include: P. Pearson and T. O'Garra (2005) 'A way forward for urban transport?' *Public Service Review: transport, local environment and the regions*, PSCA International, issue 6; T. O'Garra (2005) 'Comparative analysis of the impact of the Hydrogen bus trials on public awareness, attitudes and preferences: a comparative study of four cities' Final Project Report, AcceptH2 – funded by European Commission 5th Framework Programme Contract ENK5-CT-2002 80653: www.accepth2.com; and T. O'Garra, S. Mourato, P. Pearson (2005) 'Analysing awareness and acceptability of hydrogen vehicles: a London case study', *International Journal of Hydrogen Energy*, 30, 649–659.

Professor Peter Pearson is Professor of Energy and Environmental Studies, and Director of the Imperial College (London) Centre for Energy Policy and Technology (ICEPT), and also is a director of Imperial College's 'Futures Lab'. His research interests concern long-run fuel and energy technology transitions and their energy and environmental policy implications; and sustainable innovation policy processes. He is principal investigator of the UK Energy Research Centre's *Technology and Policy Assessment Function*, and was principal investigator of the EPSRC project on hydrogen infrastructure for vehicle refuelling in London. He is chair of the International Evaluation Panel on environment and societal sciences for the Research Council for Biosciences and Environment, Academy of Finland; and also a member of the European Commission Advisory Group on Energy for the 6th Framework RTD Programme. His many publications include: Fouquet, R. and Pearson, P. (2003) 'Five centuries of energy prices', *World Economics*, 4, 3; Hutchinson, E.J. and Pearson, P. (2004) 'An evaluation of the environmental and health effects of vehicle exhaust catalysts in the UK', *Environmental Health Perspectives*, 112, 2, 132–141; and Foxon, T.J. and Pearson, P. (2006), 'Policy processes for Low carbon Innovation in the UK: successes, failures and lessons', *Energy Policy*.

Dr. Miriam Ricci is a Research Fellow in the Institute for Social, Cultural and Policy Research, University of Salford, UK. She was a member of the EPSRC-funded UK Sustainable Hydrogen Energy Consortium, carrying out analysis of scientific expert risk assessment and also studies of public perceptions of risk associated with hydrogen energy. Previously, she worked in the electricity industry, and for the European Parliament, and she has held a Marie Curie doctoral fellowship at the Institute of Innovation Research, University of Manchester. Her main research interests are science, technology and innovation policy and governance; renewable energy; the hydrogen economy; and public engagement in science and technology. Publications include: M. Ricci *et al.* (2006) 'Hydrogen: too dangerous to base our future on?' *Hazards XIX: Process Safety and Environmental Protection*, Institution of Chemical Engineers; M. Ricci *et al.* (2006), 'Hydrogen: a matter of perception?' *The Chemical Engineer*, November, 785, 22–23; and M. Ricci *et al.* (2007) 'The transition to hydrogen-based energy: combining technology risk assessments and lay perspectives', *International Journal of Energy Sector Management*.

Fionnguala Sherry-Brennan is a Doctoral candidate at the University of Manchester, UK investigating social representations of hydrogen. Previously she obtained a Master's degree in biodiversity and conservation.

Dr. Vicky Simpson was a Research Associate, and is currently Centre Administrator, at the Centre for Sustainable Urban and Regional Futures (SURF), University of Salford, UK. Her research interests are in social studies of technology and, in particular, in relationships between sociotechnical change and organisational cultures. Her research also addresses the public understanding of science with a specific focus on public engagement and renewable energy technologies, and issues of knowledge transfer between higher education and other organisational contexts.

Dr. Lada Timotijevic is a Research Fellow in the Food, Consumer Behaviour and Health Research Centre, University of Surrey, UK. Her research interests include: risk communication; public participation and governance; identity processes and behavioural change; adaptation to rapidly changing environments (e.g. migration). Publications include: Barnett, J., Timotijevic, L., Shepherd, R. and Senior, V. (in press): 'Public responses to precautionary information from the department of health (UK) about possible health risks from mobile phones', *Health Policy*; Timotijevic, L. and Barnett, J. (2006) 'Managing the possible health risks of mobile phones: Public understandings of precautionary advice and action', *Health, Risk and Society*, Vol. 8(2), 143–164; Timotijevic, L. and Raats, M. (2006): 'Evaluation of two methods of deliberative participation of older people in food policy development', *Health Policy* doi:10.1016/j.healthpol.2006.09.005; Timotijevic, L. and Breakwell, G.M. (2000): Migration and Threats to Identity, in the Special Issue: Social changes in globalised societies and redefinition of identities: Social psychological perspectives. *Journal of Community and Applied Social Psychology*, 10: 355–372.

1
Risk and the Public Acceptance of New Technologies

Rob Flynn

Our everyday world is constantly changing. Economic growth, globalisation and the continuous development of technology ensure that in all aspects of our lives – foodstuffs, energy, clothing, transport, health, employment, leisure, etc – established practices and equipment rapidly become obsolete and there are relentless pressures to innovate and 'modernise'. In the majority of cases, these processes are driven by the capitalist market, as producers seek to shape consumer demand and as entrepreneurs champion new products and services. Part of the in-built dynamic of modern capitalism is that it is a profit-driven 'growth machine' characterised by perpetual technological innovation (Saunders, 1995). Such innovation consists of attempts to minimise or avoid technical problems with current machines and systems, and/or to increase cost efficiency, and/or to achieve radical breakthroughs to introduce completely new devices and methods. The shifts from coal to steam power, the availability of electricity, the evolution of motor vehicles and then air transport, the adoption of nuclear energy, the advance of computers and digital telecommunications, and new biotechnology industries are only a few illustrations of the fundamental transformations which have occurred in a relatively short historical period.

In contemporary western societies, if something is described as 'hi-tech' it is regarded as desirable, fashionable, a symbol of progress. Terms such as 'advanced', 'new' or 'state-of-the-art' are frequently attached to objects and machines in large-scale commercial advertising campaigns, and mass media coverage highlights their novel features. In the private sector of the economy, huge amounts of resources are devoted to research and development, and then to persuading consumers to adopt the resulting products. In recent times, perhaps the

1

speed with which mobile telephones spread, and their now almost-constant updating and increased sophistication, indicates the scale and rapidity of technological diffusion. Governments, of course, play an important role in enabling economic conditions for growth and technological development, and in many cases, have systematic strategies to encourage and subsidise industrial innovation and scientific research, linked with both civil and military applications. Nevertheless, it is usually the case that the primary motor for technological innovation – because it is so dependent on massive investment and risk-taking – is competition in the private sector.

These perhaps commonplace observations are essential for providing a context within which to assess the ways in which citizens and consumers are involved in decisions or policies about the acceptability of new technologies. It is a matter of debate as to whether members of the public are genuinely able to express their views about – let alone their approval of, or willingness to accept – a new technology in advance of the introduction of such technology. Conventionally, manufacturers, retailers and service-providers will carry out extensive market research before launching a product or service, using surveys and focus groups for example. However, technological systems and the complex infrastructure on which they depend evolve at different rates and are at such a scale that it may not be feasible to imagine consumers being asked to approve their implementation. Moreover, even though the ultimate 'test' of acceptability is consumers' willingness to pay, in their purchasing decisions consumers may 'trade-off' price against other criteria. Although many products – fuel, food, clothing, motor vehicles, for example – may in most respects be generically identical, producers attempt to differentiate their own product from others and, through advertising, claim distinctive properties for it. Consumers may associate certain products, or companies supplying them, as more prestigious, reliable or trustworthy, or as affording greater value-for-money; others are prepared to pay higher prices for designer-labels as they confer status, while others (perhaps a growing minority) seek out 'organic' foodstuffs, or 'fair-trade' products, or only support companies with 'ethical' investment portfolios, or use energy firms which supply 'green' electricity.

Consumption patterns and expenditure decisions are not determined by rational economic calculations alone, but are mediated by cultural and political factors, and are influenced by the mass media. Customer preferences and satisfaction have many different elements, but consumer demand is not usually driven by public consultation or referenda. As Clarke and Short (1993: 381) cogently noted, the major influence on policy comes from organised interest groups, 'not from an undifferentiated "public"', and elites and professional experts 'are the main institutional actors who make choices among technologies (including neglecting to develop them)'.

Why are these considerations relevant to the debate about the public acceptance of new technologies? First because much of the conventional debate tends to treat citizens as acting differently from consumers of commodities and services, and it is necessary to challenge the taken-for-granted notion of acceptability. Second because many of the most significant new technologies (for example, Genetically Modified Organisms, Nanotechnology, Hydrogen energy) have such far-reaching implications that they are seen as having unpredictable or highly contentious outcomes for the wider public interest. Consequently they are not solely private matters for the market. Third, because many of the claims being made for these new technologies are being advanced and endorsed by powerful groups of experts whose assumed scientific authority is increasingly being questioned. Populations are being invited (or persuaded) to take risks with technologies whose effects are extremely uncertain, and are also being expected to trust information and advice from governments and scientists after a period of turbulence associated with scares over medical and ecological controversies, often characterised as the 'risk society'.

This chapter therefore sets the scene for the contributions which follow, first by highlighting some of the most salient arguments about risk and risk society; second by reviewing arguments about whether and how 'the' public can be involved in shaping policy on new technologies; and finally it indicates how these issues connect with the themes taken up by the authors.

Risks in the risk society

The social scientific literature on risk and the so-called risk society is now so vast that it is impossible to provide a comprehensive overview. For several decades, scholars in Europe and the USA, from numerous disciplines, have identified the increased prevalence and impact of all types of risk as a special feature of late modern society, and as requiring new sub-disciplines, academic and technical journals, and professions dedicated to risk evaluation, risk perception and risk management. Krimsky and Golding (1992) collected a wide-ranging series of essays by leading authorities in the field, and at that time noted that risk studies had developed out of the need to regulate technology and to protect citizens from natural and technological hazards. From a more sociological theoretical standpoint, Giddens, in a number of publications but perhaps most importantly in *Modernity and Self-Identity* (1991), argued that modernity itself is a risk culture. The concept of risk has become an organising concept for lay actors as well as experts; people have become more reflexive, questioning risks but also embracing (some of) them, while all have become dependent on trust in abstract systems. Giddens (1998: 25) noted that new technologies increasingly 'penetrate' the core of our lives, yet at the same time in risk society 'we increasingly live on a high technological frontier which absolutely no one completely understands, and which generates a diversity of possible futures'. Ulrich Beck's *Risk Society* (1992; original 1986) echoed Giddens' emphasis on reflexive modernisation and introduced the concept of the 'risk society' to denote a special stage in which the nature of hazards, and their lack of temporal and spatial limits, are so profoundly different from previous eras that they are creating a new inequality, as well as a crisis of credibility and trust. These risks are 'man-made' hybrids, combining cultural definitions and technologies, politics and mass media. But, argued Beck, 'we no longer choose to take risks, we have them thrust upon us' (Beck, 1998: 12). Thus for Beck (2000), risk discourse begins where trust in security, and belief in progress, ends, and where uncertainties are manufactured. Paradoxically, science and technology solved major problems, but also (albeit unintentionally) produced new hazards.

Numerous commentators have repeatedly stressed the ambiguous, contested and opaque character of the concept of risk and it is not surprising therefore that there is such an enormous scholarly and technical literature about different aspects of risk (for important commentaries and collections, see: Boyne, 2003; Lupton, 1999; Mythen and Walklate, 2006a; Taylor-Gooby and Zinn, 2006). It is acknowledged that there are different disciplinary approaches to the concept of risk, and both within and between disciplines, various conceptual frameworks or paradigms exist. While simplifying these variations is unwise, it is nevertheless evident that writers and researchers diverge most notably according to whether they adopt a 'realist' or a 'social constructionist' standpoint. The former, crudely, subscribes to the view that risks (as hazards with potentially negative effects) are real, objective, identifiable and measurable – and that some are potentially capable of being subjected to control and management. The latter viewpoint argues that what social groups perceive as risks or threats is culturally and socially-defined: risk therefore varies between groups, over time and between cultures, and is affected by individuals' meanings and values in specific contexts. Without rehearsing the etymology of 'risk' or recounting all of the nuanced and multi-layered accounts given by different theorists, it is necessary to note further some of the principal arguments.

It is instructive that Renn (1992: 54) observed that in their extreme versions, both the positivistic and social constructivist approaches to risk were 'poor descriptions of reality' yet the debate has continued unresolved. Beck (2000) emphasised that the realist *versus* constructivist dichotomy was *not* an 'either/or' option; he favoured a pragmatic position, claiming to be both a realist and a constructivist because 'Risks are simultaneously "real" and constructed by social perception ... Their reality springs from [their] impacts' (Beck, 2000: 219–220). Others have criticised this ambivalence. Wynne (1996) for example – using a sociology of scientific knowledge (SSK) framework – questioned scientistic claims to objectivity of risk, and focused on the significance of lay knowledge and perceptions. However, Adam and Van Loon (2000) pointed out that while risks are necessarily socially constructed, they are not totally imaginary and we are not free to choose (or ignore) all potential risks. Lupton (1999) distinguished 'strong' and 'weak' forms of social constructionism in relation to risk, and Tulloch and Lupton (2003) illustrated the contingent and contextualised nature of different groups' 'risk knowledges', also showing that some people relished the positive aspects of risk-taking. Mythen's (2004) critique of Beck also stressed the heterogeneity of cultural understandings of risk, and argued that 'risk situations are more diverse, complex and multi-dimensional than the risk society narrative implies' (*op cit*, 68). In the field of environmental hazards, Irwin (2001a) argued that neither a realist nor a constructivist approach was adequate by themselves, because 'neither the natural nor the social can be given paramount status ... instead a process of co-construction needs to be recognised ... [to] avoid both objectification of the natural world and social relativism' (*op cit*, 16).

More recently, Taylor-Gooby and Zinn (2006) suggest there has been an intellectual shift away from the constructivist/realist dichotomy, highlighting a number of contemporary studies which focus on the concreteness and immediacy of people's experiences in different contexts. Wilkinson (2006) too has stressed that in everyday life, individuals' perceptions of risk are more contradictory and varied than the 'realities' ascribed by risk society theory. Again, however, Mythen and

Walklate (2006b) and Flynn (2006) – endorsing a critical realist perspective – insist that we recognise that there are real risks independent of people's knowledge of them. It is sufficient to note here that these arguments – which reflect much broader (and arguably more fundamental) epistemological and theoretical debates about agency and structure, for example – are likely to continue (see Althaus, 2005; Webster, 2004; Taylor-Gooby and Zinn, 2006). Researchers on risks (as hazards) are increasingly using eclectic approaches and – as many of the following chapters illustrate – concentrate on specific institutions, processes and experiences rather than abstract models.

Public perceptions of risk and the question of trust

Having briefly noted these long-running disputes about the status of risks and the importance attached to people's attitudes and understandings, it is necessary to consider how perceptions are formed and what influence they might have on behaviour. Discussion of the social acceptability of new technologies must include some assessment of the role and effects of public perceptions. Krimsky (1992) observed that interest in studying risk perception stemmed from two major trends: first, increasing evidence of public concern about natural hazards (earthquakes, floods, hurricanes) and man-made disasters (nuclear accidents, toxic chemical pollution); and second, widespread interest in decision-making in conditions of uncertainty. At a more general level, and more recently, other writers have commented that there is in contemporary culture almost an obsession with risk, safety and security, and this has channelled attention on public awareness of, and attitudes towards, all kinds of scientific and technological developments. Furedi (1997) was especially critical of what he termed the 'explosion' of (apparent) risks, the institutionalisation of the precautionary principle, the inflation of 'danger' and the rise of a 'culture of fear'. Wilkinson (2001a) reviewed various theoretical accounts of and evidence for the growth of anxiety in a risk society, noting that the cultural climate was pervaded by a preoccupation with harm and danger (and its avoidance). Wilkinson (2001b) further stressed that people construct and experience their knowledges of risk and uncertainty in complex and sometimes contradictory ways. Elliott (2002), in a critical review of Beck's 'risk society' thesis, nevertheless acknowledged a generalised climate of risk had been mirrored by increased attention to reflexivity and ostensible searches for agency, choice and personal responsibility. From different theoretical positions, scholars have converged upon the centrality of people's beliefs and opinions about a range of ostensible risks.

Probably the most influential and longstanding body of research has evolved out of the psychometric paradigm, influenced mainly by cognitive psychology. However, it should be noted that Wildavsky and Dake (1990) examined the evidence then available and found that there was no specific 'personality structure' for risk-taking or risk aversion; perception of danger was (and is) selective and influenced by cultural factors. It was not knowledge of a technology which caused people to worry about its dangers, they argued, but the credibility of the information about it and degree of confidence invested in organisations associated with it. For Wildavsky and Dake, the most important factor influencing the perception of risks is trust or distrust, and that raised broader issues than cognitive psychology conventionally addressed.

Nevertheless, the psychometric paradigm has generated a massive number of studies and has been extremely influential in risk perception research. One the leading researchers in this field, Paul Slovic (1992) explained that their objective was to measure, on different scales, the degree to which people were aware of specific risks, and to quantify the positive and negative criteria or values associated with them. For Slovic and colleagues, their focus was upon the *subjective* appreciation of risk. They proposed different models (comparing known with unknown risks, and the degree of 'dread' implied by a possible hazard) while accepting that explanations were multi-factorial and required multi-disciplinary approaches.

In a synoptic overview of Slovic and his colleagues' research programme (Slovic, 2000), a number of empirical patterns were identified. First, and most importantly, the concept of risk meant different things to different people, and laypeople's judgements differed from those of experts. Second, risk perceptions were partly dependent on people's intuitive and experiential thinking, and were affected by 'affective' or emotional factors. Third, where risks and benefits were defined, perceived risks declined as perceived benefits increased. Fourth, they found that trust was a crucial variable influencing perception and the communication of information; laypeople's judgements about risks described by experts and scientists were conditioned by the degree of trust in the source of information, and this varied between groups and across different issues. Sixth – and this is relevant for the discussions in later chapters – people were willing to tolerate or accept higher levels of (perceived) risk if the processes involved were voluntary, immediate, known or familiar, and were seen as controllable. Their work also indicated that risks could be subject to 'social amplification' through the media and other aspects of the communication of information, suggesting that people's responses to different messages or 'signals' may themselves intensify or magnify the image of a potential hazard.

This 'social amplification of risk framework' (SARF) has also generated a substantial research programme, which cannot be detailed here. Its central claims are, however, important and relevant for our understanding of what factors influence the acceptability of a technology. Flynn, J. *et al.* (2001) addressed the question of why certain technologies elicit fear or dread, and focused upon the process of 'stigmatisation' in which some technologies, products and places become characterised by notions of danger, risk and impurity. Negative imagery linked with events such as accidents, contamination or pollution surfaces in media reporting, and this then 'amplifies' public perceptions. In the same volume, Kasperson *et al.* (2001) applied the SARF and argued that stigmatisation and amplification are especially linked with risks that are new, involuntary, and regarded as potentially catastrophic; they again stressed the importance of public trust or distrust in the communication of information about such risks. Frewer *et al.* (2002) showed how in addition to the social amplification of risk linked with Bovine Spongiform Encephalopathy (BSE) in Britain, media reporting of genetically modified (GM) food also increased negative attitudes; when reporting reduced, there was a lowering of perceived risk.

Reviewing the value of SARF, Pidgeon *et al.* (2003: 2) noted that it focuses upon 'how risk and risk events interact with psychological, social, institutional and cultural processes in ways that amplify or attenuate risk perceptions and concerns and thereby shape risk behaviour, influence institutional processes, and affect risk consequences'. They pointed out that while trust is a crucial concept, it is not a unidimensional variable. Kasperson *et al.* (2003) further noted that in the communication of risks and risk events, both are portrayed in different images or symbols, which then interact with various cultural, psychological and institutional processes to increase or decrease the perception of risk. They too, while accepting that social trust is an important factor in these processes, regarded trust as highly variegated and poorly understood by researchers.

It is generally understood that people's attitudes towards risk, and their degree of trust in technologies and the experts who explain them, are strongly influenced by what various writers term social trust

(Siegrist and Cvetkovich, 2000). This means that there are dimensions of trust which reflect a person's or group's wider social positions, normative values and experience, and which extend beyond the immediate risk in question. For example, it has been shown that people who trust scientists and commercial companies generally are more positive about new technology. Those who trust agencies and the managers who run a facility perceive fewer risks, or lower levels of hazard, than people who are less trusting. Siegrist (2000) showed that in genetic modification technology, people's trust in the organisations or scientists carrying out the research (or using products) was the most important factor influencing their perceptions. Acceptance of the technology was directly determined by perceived risk *and* benefit, but trust had an indirect effect on the acceptance of this technology.

In one very detailed recent study of attitudes towards nine different technologies and potential hazards, Siegrist *et al.* (2005) demonstrated that for both technological and non-technological risks, higher levels of general trust and confidence reduced the perceived risk. But Poortinga and Pidgeon (2005) raised the question of whether trust is the determining factor in public perceptions and acceptability of risks. In their large-scale quantitative study of perceptions of the regulation of GM food in the UK, Poortinga and Pidgeon found that rather than being the determinant factor, trust was an expression or indicator of the acceptability of GM food. They argued that in relation to specific risk estimations (about particular technologies, for example), people's assessments are shaped by more general 'evaluative judgements'. They also suggested that trying to increase trust by simply providing (more) information to people – as a way of improving the communication of risk and its regulation – might be construed as failing to take the public's concerns seriously, and may even be counter-productive. Attempts by governments or institutions to enhance general levels of trust among people in order to secure greater acceptance of a controversial technology may be ineffective, unless they also address their specific concerns about a particular risk issue. Moreover, as Poortinga and Pidgeon (2004) observed, people holding strong preconceived beliefs may not be willing to change their views about trust, as they tend to interpret claims and evidence according to their prior values.

The asymmetry of expertise

Whatever their core beliefs and values, most laypeople are very – or wholly – dependent for information about new technologies, their risks and benefits, on scientific experts and regulatory agencies. Accepting and using a new technology implicitly means believing and trusting assurances from official bodies that those technologies are desirable, feasible and worthwhile, and that they pose no significant harm to users or threat to public safety. But there is an informational asymmetry which confers significant power on those claiming authority over the basis on which risk assessments and risk management are undertaken. There has been extensive and wide-ranging debate about the problems this creates for the public understanding of science, which again can only be referred to briefly here.

Irwin and Wynne (1996) directly challenged the orthodox assumption then found amongst most policy-makers and scientists that the general public lacked sufficient knowledge and understanding of basic science and technology. This dominant 'deficit model' implied that if citizens were better informed, and more 'scientifically literate', they would be able to make more rational decisions. Irwin and Wynne criticised the 'PUS' ('public understanding of science') model for its axioms that science was objective and benign, and that the ignorant public needed to be educated about the facts. They pointed out that science was itself socially and politically mediated in different ways, and the knowledge produced was contestable and contested. Irwin and Wynne also importantly remarked that there are many different 'publics', located in different contexts with different knowledges – a point we shall return to below, and is taken up by several later chapters. Wynne (1996) particularly argued that lay knowledge had its own validity and that scientists' credibility was problematic.

Official governmental acknowledgement of wide public disquiet in Britain prompted a review of science communication and public consultation (see Office of Science and Technology and Wellcome Trust, 2001). This signalled the need for a shift away from the deficit model towards a two-way dialogue between citizens and scientists: this was described as an 'engagement model'. Irwin (2001b) examined several exercises in public consultation in Britain about developments in biosciences, following the BSE crisis and controversy over GM food, and identified some limited movement towards greater transparency and acknowledgement of scientific uncertainty, and the possible emergence of the 'scientific citizen'.

However, Irwin and Michael (2003) were sceptical about whether there had been any significant evolution of public engagement, and even detected the continuation of the deficit model in some policy areas. They were critical of policy-makers' assumption that increased awareness and knowledge would promote stronger public support for scientific and technological innovation, again re-emphasising that people routinely make judgements in terms of credibility, trustworthiness and usefulness – but all conditional upon their own social circumstances and cultural identities.

More recently there have been a number of other efforts by government and scientific bodies to involve the general public in debate about alternative futures in science and technology innovation, but it is still arguable that these cannot overcome the inherent imbalance in knowledge and expertise. They also do not directly confront the 'framing' of such debates within certain assumed parameters; frequently public consultation presents limited options or choices, and requires citizens to express preferences in selecting among restricted priorities without permitting a much more fundamental questioning of wider goals and objectives and their desirability. Conventional techniques to elicit public opinion have also relied upon large-scale questionnaire surveys, which themselves are open to methodological criticism when investigating unknown or emergent technologies, and may fail to reveal the complexity and specificity of local, contextualised, lay knowledge. However, recently there has been a movement to experiment with other methods, and some growing official acknowledgement of the value of carrying out so-called 'upstream engagement'.

Upstream engagement and the acceptance of new technologies?

Questions about whether and how far citizens can be involved in major policy decisions – separate from the conventional electoral process – have vexed governments in many western countries over many years. Consultation and public participation are regarded as necessary and beneficial to secure the legitimacy of governmental decisions in some, but not all, policy areas. Involving the public in some aspects of risk analysis and management has been promoted as worthwhile by policy-makers in both Europe and North America since the 1990s. However, there are a number of recurrent problems noted by various commentators: which members of the public can and should be involved, what does consultation consist of, and how in representative democracies can this be reconciled with the government's responsibility to make authoritative decisions in the collective interest? These fundamental questions persist, and are especially relevant in assessing the public acceptability of new technologies.

Examining the mechanisms for identifying public concerns and preferences in risk assessment, Renn (1999) noted that there are various important factors which constrain the ability of policy-makers (and citizens) to incorporate their views in decision-making. First there is often a mistaken assumption that a single homogeneous 'public' can be identified and consulted. In practice, populations are highly differentiated locally, regionally and nationally, and there are conflicting interests and values. Consultation or participation methods have to recognise and respond to the fact that there are many different 'publics'. Second there are difficulties in deciding the most appropriate procedure for gauging public opinion – opinion survey questionnaires, face-to-face interviews, public meetings, focus groups? Different methods may yield different conclusions. Third, how can citizens be equipped to scrutinise and if necessary challenge the information provided by experts, and do they have the chance to question such experts directly? The issue of trust, and lay versus expert knowledges, remains problematic. Renn (1999) argued positively in favour of adopting what he termed 'analytic-deliberative' methods, based on co-operative discourse. This, he argued, was a 'hybrid' model of citizen participation with several stages: asking all relevant stakeholder groups to identify their own values and criteria for judging different policy options; experts from different perspectives evaluating the relative performance of alternative options; then inviting a randomly selected set of citizens panels or juries to question witnesses (from experts and interest groups), to debate and 'deliberate' on a preferred solution or decision.

Other commentators have reviewed some of these methods, and focused on other practical limitations. Thus, for example, Rowe and Frewer (2000) examined different techniques for involving the public in science and technology policy, and asked whether 'involvement' meant communication of information, or participation in decisions. Their inventory of procedures included methods of eliciting opinions (public opinion surveys and focus groups) and methods of eliciting judgements about specific decisions (such as consensus conferences and citizens' juries). They compared various techniques in terms of criteria such as the representativeness of the groups being involved; whether the process was facilitated independently to avoid (or minimise) stakeholder bias; whether citizens were involved at the earliest stage of discussion of policy; and whether the outcome of the consultation would have a genuine impact on the ultimate decision, etc. Rowe and Frewer's evaluation indicated that each method had advantages and disadvantages, no single method could be regarded as optimal, and their recommendation was for using a mixture of methods.

Similar problems and constraints were observed in the UK, following the controversy over GM crops and foods. Thus for example, Grove-White *et al.* (2000) criticised the conventional 'one-way' provision of information to the public about new technologies as being completely inadequate. A central feature of their study was the apparent reluctance of science experts, policy-makers and industrial stakeholders to acknowledge questions of scientific uncertainty. They also found that different expert stakeholders held different views about which 'facts' to provide as information to the public. Moreover, they found some unwillingness among experts to see the process of engaging with the public as a reciprocal or two-way process. Their qualitative research showed that ordinary members of the public exhibited considerable scepticism, and this critically influenced the degree of trust afforded by them as citizens rather than as consumers. Most importantly, Grove-White *et al.* (2000) concluded that people do not simply 'decide' or 'make choices' about technological innovation in the rational way usually assumed by policy-makers; in many instances their judgements are shaped by significant others, their own experience and wider aspects of trust. They concluded that, in relation to the process of 'engaging' the public, there needed to be a shift from an 'information' model to an 'interactive understanding' model. Dialogue should comprise mutual learning, they argued.

Other researchers have questioned not just whether and how dialogue and 'engagement' occurs, but also whether it has any discernible impact on technological decisions. Rowe *et al.* (2005) examined evidence about the UK 'GM Nation' public debate about the commercialisation of transgenic crops. They noted that it is difficult to specify criteria for evaluating the quality and effectiveness of different engagement processes, but argued that in the 'GM Nation' consultation exercise, the participants were not representative of the wider public; that resources for the exercise were inadequate; and that the ultimate influence on policy was unclear or minimal. These findings again raise questions about the purpose and objective of public engagement – is it substantive, resulting in changes in policy, or is it symbolic, designed to pre-empt potential complaints about lack of consultation? Some aspects of these questions are taken up in greater depth in the chapters below, but they raise normative, not just analytical, issues about what it is to be an active citizen and different models of citizenship.

Some writers have observed that conventional models of participation are framed within a liberal, pluralist framework, premised on individualistic notions of the citizen and a representative democratic system, which begs questions about the distribution of power. For example, in a critique of what they termed the 'uncritical enthusiasm' for deliberative methods, Leach *et al.* (2005) pointed out that even if citizens' juries and deliberative panels are used, it is unclear whether they are passively reacting to an agenda already established by experts and stakeholders, or can influence the agenda itself. Leach *et al.* (2005) argued that public engagement exercises are framed within the dominant scientific discourse and mainstream problem-definitions, and imply that citizen influence is marginal. In the same volume, Wynne (2005) was also critical of the extravagant optimism associated with the adoption of deliberative methods of engagement. He argued that many citizen participation exercises focused exclusively on 'downstream' risk and impacts. The assumption of those commissioning such exercises is that members of the public are mainly interested in the effects of a technology, rather than the 'upstream' motives and objectives underlying the innovation. This position, he further argues, reflects a belief that the definition of the issue to be consulted on is ascribed by experts. Wynne suggests that the recent popularity of citizen engagement in science and technology is a 'mirage'.

Somewhat similar views have been advanced by Stirling, one of the leading proponents of alternative means of facilitating authentic public involvement in 'upstream' assessment of technology. Stirling (2005) asked whether public participation or engagement methods 'opened up' or 'closed down' effective debate about innovation. He contrasted orthodox expert-dominated procedures (such as risk assessment, Delphi methods) with more recent deliberative techniques (such as citizens' panels, consensus conferences) but radically questioned whether either approach opened up or closed down debates. 'Closing down' implies that the method is instrumentally seeking a justification for an authoritative decision, framed within very specific and limited parameters. By contrast, 'opening up' the process of technology choices entails posing alternative possibilities, acknowledges scientific uncertainties and incorporates what might otherwise be seen as marginal perspectives, to explore a range of solutions. Here then, in both cases, the methods or procedures for involving the public are secondary concerns: the principal issue is the framing and purpose of the engagement.

This emphasis on specifying the scope and impact of public engagement through 'upstream' consultation was further developed by

Wilsdon and Willis (2004). Their starting point again was the apparently low levels of public trust in government and their scepticism about influencing decision-making in science and technology innovation. They noted that moving public engagement 'upstream' has begun to enter the lexicon of science policy-making only relatively recently. They suggest that governments have slowly accepted – and only to a limited extent – the virtue of enabling public debate at early stages of the research and development process, rather than much later (and 'downstream') when the technologies are already developed and waiting to be applied and implemented. They note that in most cases in government and industry, attempts at dialogue with the public about new innovations happen (if at all) much later in the business process when strategic decisions have already been made. This they believe, is counterproductive, and may further undermine public confidence and trust.

For Wilsdon and Willis (2004: 24) the purpose of upstream engagement is: 'to make visible the invisible, to expose to public scrutiny the values, visions and assumptions that usually lie hidden'. In practice, they propose, policy-makers and stakeholders should engage with the public and directly respond to citizens' questions such as: 'Why this technology? Why not another? Who needs it? Who is controlling it? Who benefits from it? Can they be trusted? What will it mean for me and my family?' (*op cit*, 28). As they emphasise when posing these questions, just because a technology is possible or feasible does not mean that it is desirable, or desired. So, instead of dealing only with possible risks, upstream engagement asks questions about the very desirability of the technology and whose goals, values and interests it might serve. Thus, asking people to give their views about the 'acceptability' of a technology is not merely a process of eliciting their preferences for different technical features of a product or process, or their perception of risks, but entails recognition that there are normative choices and political priorities to be established from the outset of the debate.

'Acceptability' or acceptance of what?

The very notion of 'acceptable risk' has been current for many decades in the conventional risk assessment literature and has been the subject of various criticisms. Otway (1992) commented that risks are never 'accepted' in the abstract; people judge possible risks and benefits in relation to specific, immediate or anticipated effects on their everyday lives. But Otway also argues that narrow concepts of risk perception and 'acceptance' often obscure proper recognition of the *totality* of the system and technology under consideration. Wynne (1992) strongly criticised the 'acceptable risk paradigm' because it misleadingly simplifies risks on reified quantitative scales, assumes a singular meaning for risk, and fails to apprehend the varying contexts and meanings found among laypeople. Later, other writers challenged the idea that citizens' or consumers' choices (in relation to technological risks) were entirely voluntary. For example, Purcell *et al.* (2000) noted that unconstrained choices do not occur, and that the 'menus' of choice are limited and are institutionally constrained. People are unable to make systematic evaluations of risks and benefits without considerable technical knowledge to allow discrimination between possible options. As Purcell *et al.* (2000: 67) also comment, 'Individuals cannot "choose" and thereby consent to risks they do not understand'. Further, people 'choose' – or accept from – 'among a highly truncated ... menu of products and technologies that have made it through the production process' (*op cit*, 73).

Nevertheless, the conventional risk management literature persists in approaching these issues in a technocratic way. Klinke and Renn (2002) in a critical review, showed how risk management is essentially about reducing risks to a level deemed 'acceptable' by society. They defined risk evaluation as a process by which individuals, agencies and social groups determine the acceptability (or 'tolerable level') of any given risk; if such risks are viewed as unacceptable, then, in theory, appropriate organisations seek means of reducing the risk accordingly, through risk management. However, Klinke and Renn suggest a different approach to these issues, favouring analytic-deliberative methods. Other researchers have examined detailed case studies of public reactions to proposals to locate or develop controversial facilities or technologies in their locality, and raise important questions about public acceptability.

For example, Wolfe *et al.* (2002) studied responses to plans to develop hazardous waste remediation technologies in the USA. They devised a conceptual framework for analysing the public acceptance of controversial technologies (PACT). They distinguished between public *acceptability* and public *acceptance*; the former refers to people's willingness to consider the technology seriously, the latter means the formal decision to implement the proposal. They considered different dimensions of this in terms of legitimacy, representation, exclusion and power and authority. For Wolfe *et al.* 'acceptability' deals with the extent to which the technology conforms with social values and norms 'sufficiently well to be placed on the table as a viable alternative to other technologies' (*op cit*, 137). They emphasise that: a technology might be technically feasible but not meet the test of social acceptability; acceptability is a *continuum* not a dichotomy; and that acceptability may change over time, positively and negatively. Space prevents an extended discussion of other studies and their approach to what constitutes acceptability or acceptance, but it is evident that acceptability or acceptance are both value-laden terms with different dimensions, and have become politicised concepts.

If we consider in general terms some of those dimensions, a number of observations can be made. Acceptability might be interpreted from different standpoints. From a citizen/consumer viewpoint, acceptability might signify positive approval, consent and active endorsement, but it might otherwise signify acquiescence, resignation and passive compliance because rejection or resistance is infeasible. From the standpoint of the advocate or promoter of a new technology, acceptability or acceptance of any description might be sought, although it can be assumed that the more active form might be preferred by stakeholders because it appears to confer legitimacy. In everyday life, consumers may not even be consciously aware of a decision to voluntarily approve or accept a technology – many, perhaps most, aspects of our technologies are simply there, mundane and taken-for-granted, and there may even be indifference to some of them. On the other hand, in the private market, commercial advertising and the promotion of new products, depend on 'enrolling' consumers in a technology as part of a lifestyle choice; such consumers may exhibit aspirational purchasing, and then, as enthusiasts, willingly extol the benefits of such products.

The factors influencing either position are multiple and varied. Either as citizens or consumers, those who voluntarily accept a new technology may do so for various reasons – trust in information and advice from experts or scientists, recommendation by government agencies, mass media reporting and advertising, and family and peer-group pressure. Conversely, some people may not have made any choice or decision about whether to 'accept' a technology, firstly because no such choice or decision was ever offered to them, or because of coercion of circumstances or state regulation, there are no alternatives. Technological obsolescence – or claims of improved efficiency and cost – may have been used to justify the shift from coal gas to natural gas, for example, but few people had much choice in the matter. The planned switchover to digital television broadcasting will also compel consumers to re-equip, irrespective of their preferences. Similarly, the introduction of biometric passports in the UK and USA is effectively compulsory, and has not been the subject of public consultation, even though the consequences for surveillance and privacy are contentious. Thus 'acceptability' or 'acceptance' cannot be treated as unproblematically positive.

It is also important to note that acceptance or acceptability, of whatever kind, can be expected to be differentiated by socio-economic group, age, gender, perhaps ethnicity and region – much depends on the existing and historical cultural context in which the technologies and risks are being considered. People's willingness to accept risks, and to trust agencies and companies claiming to manage them, will vary, and their response will be subject to mediation and change. Later chapters provide some illustrations of this in connection with different technologies.

One other connected issue is particularly relevant and can be considered here very briefly: even if people indicate their generalised acceptance of a proposed technology, this does not entail an open-ended commitment, and may not result in complete changes in behaviour. It is already known for example that there is a disjunction between people's awareness of global warming and climate change, and their willingness to adjust their behaviour to cope with environmental problems and pressure to adopt more sustainable lifestyles. Various researchers have shown that government policy to persuade people to reduce their consumption and practice more energy-efficient lifestyles has not been effective. There are many complex reasons for this, but as Hobson (2001) showed, the so-called 'barriers to action' and gap between awareness and action, are deeply embedded in routine behaviour and values, and not easily modified.

From a governmental viewpoint this has become regarded as an increasingly serious and urgent problem, and both academic and public bodies are directing much attention to it. For example, Jackson (2005) has comprehensively reviewed the vast literature on consumer behaviour and behavioural change. He pointed out that far from exercising deliberate choice, most people find that most of the time they are 'locked-in' to consumption patterns which are ultimately unsustainable. In the face of such evidence, the UK government has adopted a policy of sustainable development (HM Government, 2005) which recognises the need to overcome the separation between awareness and action, and advocates a series of initiatives to 'enable, encourage, engage and exemplify' sustainability and to help people 'make better choices'. The Sustainable Development Commission (2006) has taken this further, in identifying means of shifting opinions and changing behaviours, by using a deliberative consumer forum. The Sustainable Development Commission (2006) in an extensive report, also suggested a series of measures to encourage citizens and consumers, and companies and government agencies, to adopt more sustainable lifestyles and to overcome the 'I will if you will' mentality of most people. All of these reports now acknowledge that merely providing information, or cataloguing risks, is unlikely to change behaviour by themselves. Instead, policy-makers are showing some recognition of the culturally-mediated character of citizen and consumer decisions, and the necessity to use more 'deliberative' processes to open up debate about the practicalities of change.

Conclusion and chapter themes

It is clear that from a social scientific perspective, what constitutes risk, and whether such risks are subject to public scrutiny, is a still a matter for debate. New technologies may evolve in response to a variety of demands and pressures, and some may incorporate consumer market research while others may require greater citizen consultation. Public awareness of, and 'engagement' in, any of these developments is likely to vary, and whether such involvement affects the rate of innovation or influences its eventual form is far from clear. Acceptability or acceptance, as we have seen, has different dimensions, and there are many different means for seeking – or securing – it, which all raise important normative and political questions about trust in science and technology, and the public's dependency on expert knowledge.

These issues are all addressed in the following chapters. The contributors and topics were selected because they are concerned with very new and emerging technologies with uncertain but potentially far-reaching implications for all our lives. With two exceptions (Chapter 3 and Chapter 11) these chapters were developed from papers originally presented at a seminar organised by Flynn and Bellaby in an ESRC Seminar Series about 'Analysing Social Dimensions of Emerging Hydrogen Economies', a series co-ordinated by Hodson and Marvin in 2005 – (see www.surf.salford.ac.uk/HydrogenEconomies/home.htm). The editors are grateful to all the contributors and to other participants in the seminar for their eagerness to compare approaches and findings from different policy areas and to explore the scope for inter-disciplinary working. In Chapter 2, Alan Irwin, an internationally-recognised scholar in the sociology of science and technology, directly challenges the widespread belief that there has been a movement away from the deficit model towards more dialogue with the public, and discusses evidence from a recent European Union study. In Chapter 3, Tom Horlick-Jones, another leading writer on risk and new technology, emphasises that particular technologies have their own distinctive materiality and 'risk signature', and that this then is reflected in different modes of understanding, which may affect the acceptability of technologies. Chapter 4 by Fischer and Frewer provides detailed evidence (mainly from the Netherlands) about consumer attitudes towards GM foods and shows clearly the continuing importance of questions about public trust and distrust. In Chapter 5, Barnett and Timotijevic discuss the development of mobile telecommunications technologies, and consider the tensions between approaches to risk management and communication which highlight precaution, and the public's response to uncertainty. Perhaps one of the most recent innovative technologies – nanotechnology – is considered in Chapter 6, where Mohr argues that a technocratic, expert-dominated discourse dominates despite some attempts at 'upstream' public engagement. The next four chapters all have a common theme, as they are concerned with an apparent novel solution to environmental and energy crises – the so-called 'Hydrogen economy'. In Chapter 7, O'Garra, Pearson and Mourato discuss detailed studies of public acceptability of hydrogen-powered transport, vehicles and infrastructure. Chapter 8, by Sherry-Brennan, Hannah Devine-Wright and Patrick Devine-Wright, reports on a case study of a community-based renewable hydrogen energy project, and shows how the local population's attitudes were strongly affected by their specific locale and possible employment opportunities. In Chapter 9, Ricci, Bellaby and Flynn present and analyse findings from case studies of stakeholders' and the public's perceptions of the risks and benefits of hydrogen energy in three areas of the UK. The nature of opposition towards the siting of a hydrogen refuelling depot, and the different interests involved in the process of technological transition, are examined by Hodson, Marvin and Simpson in Chapter 10. How to engage the public and stakeholders in deliberation about alternative technological futures is discussed by Eames and McDowall in Chapter 11, in which they explain a recently developed procedure, 'multi-criteria mapping'. Finally, Bellaby in Chapter 12 reviews the connections and common themes between the contributions, and offers a general commentary on broader questions about risk and its public acceptability.

References

B. Adam and J. van Loon, 'Introduction: repositioning risk', in Adam, B., Beck, U. and van Loon, J. (eds) *The Risk Society and Beyond* (London: Sage, 2000).

C.E. Althaus, 'A disciplinary perspective on the epistemological status of risk', *Risk Analysis*, 25, 3 (2005) 567–588.

U. Beck, *Risk Society: towards a new modernity* (London: Sage, 1992).

U. Beck, 'Politics of risk society', in Franklin, J. (ed.) *The Politics of Risk Society* (Cambridge: Polity Press, 1998).

U. Beck, 'Risk society revisited', in Adam, B., Beck, U. and van Loon, J. (eds) *The Risk Society and Beyond* (London: Sage, 2000).

R. Boyne, *Risk* (Buckingham: Open University Press, 2003).

L. Clarke and J.F. Short, 'Social organization and risk: some current controversies', *Annual Review of Sociology*, 19 (1993) 375–399.

A. Elliott, 'Beck's sociology of risk: a critical assessment', *Sociology*, 36, 2 (2002) 293–315.

J. Flynn, P. Slovic and H. Kunreuther, 'Preface', in Flynn, J., Slovic, P. and Kunreuther, H. (eds) *Risk, Media and Stigma* (London: Earthscan Publications, 2001).

R. Flynn, 'Health and risk', in Mythen, G. and Walklate, S. *Beyond the Risk Society* (Maidenhead: Open University Press, 2006).

L.J. Frewer, S. Miles and R. Marsh, 'The media and genetically modified foods: evidence in support of social amplification of risk', *Risk Analysis*, 22, 4 (2002) 701–711.

F. Furedi, *Culture of Fear* (London: Cassell, 1997).

A. Giddens, 'Risk society', in Franklin, J. (ed.) *The Politics of Risk Society* (Cambridge: Polity Press, 1998).

A. Giddens, *Modernity and Self-Identity* (Cambridge: Polity Press, 1991).

R. Grove-White, P. Macnaghten and B. Wynne, *Wising Up: the public and new technologies*, Research Report by Centre for the Study of Environmental Change (Lancaster: Lancaster University, November 2000).

HM Government, *Securing the Future: The UK Government Sustainable Development Strategy*, Cm 6467 (London: HMSO, March 2005).

K. Hobson, 'Sustainable lifestyles: rethinking barriers and behaviour change', in Cohen, M.J. and Murphy, J. (eds) *Exploring Sustainable Consumption: environmental policy and the social sciences* (Oxford: Pergamon-Elsevier, 2001).

A. Irwin, *Sociology and the Environment* (Cambridge: Polity Press, 2001a).

A. Irwin, 'Constructing the scientific citizen: science and democracy in the biosciences', *Public Understanding of Science*, 10 (2001b) 1–18.

A. Irwin and B. Wynne, 'Introduction', in Irwin, A. and Wynne, B. (eds) *Misunderstanding Science?* (Cambridge: Cambridge University Press, 1996).

A. Irwin and M. Michael, *Science, Social Theory and Public Knowledge* (Maidenhead: Open University Press, 2003).

T. Jackson, *Motivating Sustainable Consumption: a review of the evidence on consumer behaviour and behavioural change*, Report to the Sustainable Development Research Network (Centre for Environmental Strategy: University of Surrey, 2005).

J. Kasperson, R. Kasperson, N. Pidgeon and P. Slovic, 'The social amplification of risk: assessing fifteen years of research and theory', in Pidgeon, N., Kasperson, R.

and Slovic, P. (eds) *The Social Amplification of Risk* (Cambridge: Cambridge University Press, 2003).

R. Kasperson, N. Jhaveri and J. Kasperson, 'Stigma and the social amplification of risk', in Flynn, J., Slovic, P. and Kunreuther, H. (eds) *Risk, Media and Stigma* (London: Earthscan Publications, 2001).

A. Klinke and O. Renn, 'A new approach to risk-evaluation and management: risk-based, precaution-based and discourse-based strategies', *Risk Analysis*, 22, 6 (2002) 1071–1094.

S. Krimsky and D. Golding (eds) *Social Theories of Risk* (Westport and London: Praeger, 1992).

S. Krimsky, 'The role of theory in risk studies', in Krimsky, S. and Golding, D. (eds) *Social Theories of Risk* (Westport and London: Praeger, 1992).

M. Leach, I. Scoones and B. Wynne, 'Introduction: science, citizenship and globalization', in Leach, M., Scoones, I. and Wynne, B. (eds) *Science and Citizens* (London: Zed Books, 2005).

D. Lupton, *Risk* (London: Routledge, 1999).

G. Mythen, *Ulrich Beck: a critical introduction to the risk society* (London: Pluto Press, 2004).

G. Mythen and S. Walklate, 'Introduction: thinking beyond the risk society', in Mythen, G. and Walklate, S. (eds) *Beyond the Risk Society* (Maidenhead: Open University Press, 2006a).

G. Mythen and S. Walklate, 'Conclusion: towards a holistic approach to risk and human security', in Mythen, G. and Walklate, S. (eds) *Beyond the Risk Society* (Maidenhead: Open University Press, 2006b).

Office of Science and Technology and Wellcome Trust, 'Science and the Public: a review of science communication and public attitudes to science in Britain', *Public Understanding of Science*, 10 (2001) 315–330.

H. Otway, 'Public wisdom, expert fallibility: toward a contextual theory of risk', in Krimsky, S. and Golding, D. (eds) *Social Theories of Risk* (Westport and London: Praeger, 1992).

N. Pidgeon, R. Kasperson and P. Slovic, 'Introduction', in Pidgeon, N., Kasperson, R. and Slovic, P. (eds) *The Social Amplification of Risk* (Cambridge: Cambridge University Press, 2003).

W. Poortinga and N. Pidgeon, 'Trust in risk regulation: cause or consequence of the acceptability of GM food?', *Risk Analysis*, 25, 1 (2005) 199–209.

W. Poortinga and N. Pidgeon, 'Trust, the asymmetry principle and the role of prior beliefs', *Risk Analysis*, 24, 6 (2004) 1475–1486.

K. Purcell, L. Clarke and L. Renzulli, 'Menus of choice: the social embeddedness of decisions', in Cohen, M. (ed.) *Risk in the Modern Age* (Houndmills: Macmillan, 2000).

O. Renn, 'Concepts of risk: a classification', in Krimsky, S. and Golding, D. (eds) *Social Theories of Risk* (Westport and London: Praeger, 1992).

O. Renn, 'A model for an analytic-deliberative process in risk management', *Policy Analysis*, 33, 18 (1999) 3049–3055.

G. Rowe and L. Frewer, 'Public participation methods: a framework for evaluation', *Science, Technology and Human Values*, 25, 1 (2000) 3–29.

G. Rowe, T. Horlick-Jones, J. Walls and N. Pidgeon, 'Difficulties in evaluating public engagement initiatives: reflections on an evaluation of the UK

GM Nation public debate about transgenic crops', *Public Understanding of Science*, 14 (2005) 331–352.

P. Saunders, *Capitalism: a social audit* (Buckingham: Open University Press, 1995).

M. Siegrist and G. Cvetcovich, 'Perception of hazards: the role of social trust and knowledge', *Risk Analysis*, 20, 5 (2000) 713–719.

M. Siegrist, 'The influence of trust and perceptions of risks and benefits on the acceptance of gene technology', *Risk Analysis*, 20, 2 (2000) 195–203.

M. Siegriep, H. Gutscher and T.C. Earle, 'Perception of risk: the influence of general trust and general confidence', *Journal of Risk Research*, 8, 2 (2005) 145–156.

P. Slovic, *The Perception of Risk* (London: Earthscan Publications, 2000).

P. Slovic, 'Perception of risk: reflections on the psychometric paradigm', in Krimsky, S. and Golding, D. (eds) *Social Theories of Risk* (Westport and London: Praeger, 1992).

A. Stirling, 'Opening up or closing down? Analysis, participation and power in the social appraisal of technology', in Leach, M., Scoones, I. and Wynne, B. (eds) *Science and Citizens* (London: Zed Books, 2005).

Sustainable Development Commission, *I Will if You Will: towards sustainable consumption, Report of the Sustainable Consumption Roundtable* (London: Sustainable Development Commission, May 2006). Accessed 2 May 2006 from http://www.sd-commission.org.uk/publications.php?id=367

P. Taylor-Gooby and J.O. Zinn, 'The current significance of risk', in Taylor-Gooby, P. and Zinn, J.O. (eds) *Risk in Social Science* (Oxford: Oxford University Press, 2006).

J. Tulloch and D. Lupton, *Risk and Everyday Life* (London: Sage, 2003).

A. Webster, 'State of the art: risk, science and policy', *Policy Studies*, 25, 1 (2004) 5–18.

A. Wildavsky and K. Dake, 'Theories of risk perception: who fears what and why?', *Daedalus*, 119, 4 (1990) 41–60.

I. Wilkinson, *Anxiety in a Risk Society* (London: Routledge, 2001a).

I. Wilkinson, 'Social theories of risk perception', *Current Sociology*, 49, 1 (2001b) 1–22.

I. Wilkinson, 'Psychology and risk', in Mythen, G. and Walklate, S. (eds) *Beyond the Risk Society* (Maidenhead: Open University Press, 2006).

J. Wilsdon and R. Willis, *See-through Science: why public engagement needs to move upstream* (London: Demos, 2004).

A.K. Wolfe, D.J. Bjornstad, M. Russell and N.D. Kerchner, 'A framework for analyzing dialogues over the acceptability of controversial technologies', *Science, Technology and Human Values*, 27, 1 (2002) 134–159.

B. Wynne, 'Risk and social learning: reification to engagement', in Krimsky, S. and Golding, D. (eds) *Social Theories of Risk* (Westport and London: Praeger, 1992).

B. Wynne, 'May the sheep safely graze? A reflexive view of the expert-lay knowledge divide', in Scott, L., Szerszynski, B. and Wynne, B. (eds) *Risk, Environment and Modernity* (London: Sage, 1996).

B. Wynne, 'Risk as a globalizing "democratic" discourse? Framing subjects and citizens', in Leach, M., Scoones, I. and Wynne, B. (eds) *Science and Citizens* (London: Zed Books, 2005).

2
Public Dialogue and the Scientific Citizen

Alan Irwin

> The time is ripe for government to engage earlier and more deeply with the public in the development of policies and priorities, so that they are informed by public aspirations and concerns from the outset.
>
> (Council for Science and Technology, March 2005)

> The acquisition of a basic grounding in science and technology by the European public and a regular flow of information to the public from experts are not in themselves enough to enable people to form an opinion. A true dialogue must therefore be instituted between science and society.
>
> (European Commission, 2002: 14)

In British science policy circles, talk of public engagement and dialogue has become ubiquitous. Since the late-1990s, the language of 'public understanding of science' has given way to a new emphasis on openness, transparency and re-building trust between science, government and the wider publics. Much has certainly changed in terms of the rhetoric of scientific governance since the height of the 'mad cow' (BSE) crisis. However, in this chapter I want to look more closely at the nature of this change in order to explore underlying – and often unconsidered – elements of continuity, dislocation and contradiction. In so doing, the point is not to deny the very real commitment to engagement and dialogue which exists in certain policy circles nor to diminish the very real activities that have taken place in the form of specific engagement exercises. Instead, this chapter aims to put all this 'talk about talk' into wider context and to identify some of the very significant conceptual and policy-related questions that need to be considered.

The potential tension between science-led innovation and public dialogue was well (even if unwittingly) expressed by Tony Blair in his major 'Science matters' speech to the British Royal Society in 2002. Discussing Cambridge University's plan to build a new centre for neurological research which would involve experimentation on primates, the British Prime Minister asserted: 'We cannot have vital work stifled simply because it is controversial. We need, therefore, a robust, engaging dialogue with the public. We need to re-establish trust and confidence in the way that science can demonstrate new opportunities, and offer new solutions' (Blair, 2002).

What such a formulation fails to recognise is that increased dialogue does not necessarily engender trust. Equally, it cannot simply be assumed that societal support for controversial scientific developments will be the outcome of 'robust dialogue'. Thus, the conventional formulation of 're-building trust and societal support through dialogue' fails even to consider more fundamental issues of the social desirability – or perceived social need – for socio-technical change. As this chapter will especially emphasise, it also fails to consider how 'increased dialogue' can be integrated within policy-making processes which have traditionally operated according to very different concepts of efficiency, rationality and the 'public interest' as defined by civil servants, elected officials and selected stakeholders.

In addressing these issues, it must be noted that such matters of societal engagement with technical change are not simply a challenge for policy processes but also for the social sciences. Certainly, there has been a tendency for social scientists (including those working within Science and Technology Studies (STS)) either to sloganise about such matters (usually, in terms of calls for greater democracy and openness) or to dismiss initiatives in this area as insufficient, tokenistic and legitimatory (usually with implicit reference to some democratic ideal). Of course, there is a worthy tradition within social science of advocating democratic values. It is also important for social scientists to maintain a properly critical perspective on policy developments. However, it may be that the more interesting and important questions fall outside this dichotomy between advocacy and dismissal. That at least is the working assumption behind this chapter.

Putting that social scientific point in more specific terms, the intention here is to open up a wider research agenda for the study of public engagement in scientific governance. This agenda is intended to have significant scholarly merit but also to be of relevance and importance for policy and practice in this area. As one example of this emergent agenda (this is very much work in progress), it is important to consider not simply whether public engagement is a good (pro-democracy, policy-enhancing) or bad (legitimatory, tokenistic) thing but also to explore the *compatibility* of the 'new' approach with the conventional bureaucratic and institutional culture of scientific governance.

This issue of compatibility takes at least two forms: between social experiments in dialogue/transparency and more conventional activities within science policy-making (the 'business as usual' of government departments and advisory bodies); and, between the different elements which have been clustered within the 'new' governance style (can one have transparency without dialogue or the acknowledgement of uncertainty with the deficit model?). The issue of compatibility also obliges us to ask further empirical and theoretical questions about the 'new' governance: is there evidence that greater openness will indeed encourage public trust in decision-making? Is it possible for expert scrutiny to take place within fully 'open' systems? Since the logical possibilities for this appear endless, how much uncertainty should (or can) be acknowledged within decision-making? One of the suggestions here is that normative debate over the underlying principles of dialogue and engagement has clouded such important, but more specific (and perhaps more practical), questions.

Putting all this succinctly, the argument in this chapter is that we should move from approaching public dialogue around science as essentially a matter of broad principle and instead consider it as an important site for theoretical and empirical inquiry – but also as an important test-bed for the development of innovative public policy. This chapter certainly aims to address both theoretical questions of the nature of scientific governance in the contemporary world, and pressing policy-related matters of how to enact more expansive approaches to scientific governance. In that way, it aims to encourage both scholarly quality and policy-relevance: what Pettigrew (2005) has termed the 'double hurdle' of academic research. More than that, it is written in the belief that this is not simply a hurdle but also a potential double benefit since scholarship and relevance can support and sustain one another. What follows is not intended as a full exposition of this approach but rather some tentative steps in that direction.

From deficit to dialogue?

There are many ways of recounting and explaining the emergence of a 'new' governance style in the British setting. Certainly, many converging influences have been at work. These influences include a greater emphasis on accountability, transparency and communication across many parts of the public sector (notably health and social care), an international political discussion over making government more relevant and accessible to 'the citizen' (often linked to notions of 'participatory democracy' and 'active citizenship'), and social scientific criticisms of the so-called 'deficit model' (whereby social concerns about science are seen primarily as a reflection of scientific ignorance and the need for better public education).

More particularly, science-public relations were brought to the fore by a series of political crises during the 1990s, crises which suggested a substantial difficulty for governments when dealing with issues of technical change and the public concerns that these could generate (or, more accurately, crystallise). Of course, public unease around socio-technical change was hardly new, as previous, often localised, controversies over nuclear power, environmental harm and workplace health and safety had indicated. However, the possibility of a consumer reaction against a series of food-related problems in particular brought the potential economic and innovation impacts of 'public resistance' higher up the governance agenda. Perhaps the most famous example of this emergent sense of crisis was mad cow disease (BSE).

Back in 1990 when the UK Ministry of Agriculture, Fisheries and Food (MAFF) found itself confronting a new category of risk, it acted in what was then the conventional fashion. Concerned at the prospect of an anxious or panicky public reaction (suggesting an abiding mistrust – or even fear – of the wider publics that has not gone away), and despite undeniable technical uncertainties, reassurances were offered by a variety of governmental and industrial groups – all trying to get across the message that the risks were minimal and that consumers should continue to buy British beef (or 'eat British beef with confidence' as government representatives often put it). In what is now generally viewed as a misguided attempt at media manipulation (but was intended at the time as a principled manifestation of complete personal confidence), the Minister fed a beefburger to his daughter before the world's press. As one public notice from the Meat and Livestock Commission put it:

> Eating British beef is completely safe. There is no evidence of any threat to human health caused by this animal health problem (BSE) ... This is the view of independent British and European scientists and not just the meat industry ... This view has been endorsed by the Department of Health. (*The Times*, May 18, 1990)

Whilst one can only assume that these statements were sincerely meant, it must be acknowledged that at the time of these reassurances there was already considerable uncertainty over whether this animal disease could be transmitted through beef consumption to humans. No mechanism for such a transmission had been identified and the possibility appeared remote. However, interviews at the time with government scientists suggested – at least for those working directly with such issues (at the lab bench, so to speak) – that the existence of a remote but unquantifiable risk was understood and acknowledged (i.e. it could not be anticipated but neither could it be ruled out). As one internal MAFF memo expressed it in 1988:

> We do not know where this disease came from, we do not know how it is spread and we do not know whether it can be passed to humans ... There is no evidence that people can be affected but we cannot say there is no risk. (Phillips, para 229)

Needless to say, this otherwise arcane debate generated considerable tension between government scientists who saw the risk to humans as hypothetical, unlikely and ill-defined but nevertheless impossible to dismiss, and government officials and political spokespeople who translated 'indeterminate and unquantifiable risk' into 'zero risk'. In a classic representation of the notion that 'distance lends enchantment' (Collins and Pinch, 1993), those closest to the technical evidence were of the opinion that 'absence of evidence' did not equate to 'evidence of absence'. Equally, the inability to model or quantify the threat (due to the non-identification of a causal mechanism) did not mean that it was 'zero'. Meanwhile, these 'unknown unknowns' (Grove-White, 2001) could be dismissed in policy circles as nebulous, hypothetical and over-abstracted. From this policy perspective, how can practical decision-making take account of such ill-defined possibilities – especially when, in very significant contrast, the economic impacts of greater restrictions on beef sales were likely to be immediate, tangible and all too easily quantifiable? As the official Phillips report into BSE put this some ten years later:

> The Government did not lie to the public about BSE. It believed that the risks posed by BSE to humans were remote. The Government was pre-occupied with preventing an alarmist over-reaction to BSE because it believed that the risk was remote. It is now clear that this campaign of reassurance was a mistake. When on 20 March 1996

the Government announced that BSE had probably been transmitted to humans, the public felt that they had been betrayed. Confidence in government pronouncements about risk was a further casualty of BSE. (Phillips *et al.*, 2000, Volume 1, section 1)

Terms such as 'unwarranted reassurance' (*ibid*: 1150) and 'culture of secrecy' (1258) are prominent within the Phillips report. As one witness to the inquiry put it, 'one was aware of slightly leaning into the wind ... we tended to make more reassuring sounding statements than might ideally have been said' (1294). In contrast, the Chief Scientific Adviser is quoted as arguing that the temptation should be resisted 'to hold the facts close' so that a 'simple message can be taken out into the market place'. Instead, 'the full messy process whereby scientific understanding is arrived at with all its problems has to be spilled out into the open' (1297). In this way a 'culture of trust' rather than one of secrecy can be developed. On that basis, the Phillips report reached the firm conclusion that a 'policy of openness' is the best approach to communication with the wider publics.

It is worth emphasising that the approach taken within the BSE episode was not unusual in its time. Whether concerning public information about nuclear power, the control of workplace chemicals, or wider issues of food safety, the prevailing assumption in British government circles was that the admission of uncertainty would confuse 'the public' (usually seen to be one undifferentiated mass) and only 'the experts' (narrowly defined) could handle 'the facts' in a suitably rational fashion. Expert evidence was often kept confidential since to do otherwise would invite unhelpful and uninformed interventions. In addition, confidentiality could facilitate flexibility of response so that new information might be acted upon speedily but also pragmatically. By keeping the wider publics at a distance, a more consensual and collegial style could prevail among insider experts: a style which sought legitimacy in its independence, objectivity and 'sound science' (Brickman *et al.*, 1985; Irwin, 1985).

Various reports since the late 1990s have marked the transition to a newer, more 'open' style of scientific governance. Greater transparency, recognition of uncertainty and enhanced public engagement have been advocated by a series of British reports from the late-1990s onwards (RCEP, 1998; DTI, 2000; RS/RAE, 2004; see also Irwin, 2006). As the landmark 2000 report from the House of Lords Select Committee on Science and Technology put it: 'Policy makers will find it hard to win public support on any issue with a science component, unless the public's attitudes and values are recognised, respected and weighed along with the scientific and other factors' (House of Lords, 2000: 6).

Prominent among these changes has been the development by the Government's Chief Scientific Adviser of a set of guidelines on the relationship between scientific advice and policy-making. First issued in 1997, and refined in 2000 and 2005, the key messages behind these guidelines are that government departments should publish scientific advice and all relevant papers, obtain a wide range of advice from 'the best sources' (particularly when there is scientific uncertainty) and identify the issues early when scientific advice is needed. The guidelines emphasise the need for procedures to be open and transparent but also the importance of 'bringing together the right people' – a category that might include lay members of advisory groups, consumer groups and 'other stakeholder bodies' (Office of Science and Technology, 2000).

Much then has changed in terms of the rhetoric of scientific governance since the early days of BSE. Practical initiatives also have taken place – notably around genetically modified (GM) food and stem cell research but also looking ahead to the challenges of nanotechnology. As suggested in the opening section, this has provoked a critical discussion about whether all this has 'just' been talk or rather a significant change in governance thinking (Irwin, 2006). However, this discussion has tended to be stronger on generalisation than on analysis, asking broad questions rather than considering the specific contextualities of this area of activity. Thus, to return to the BSE case, it is clear that a number of questions need to be explored with regard to these criticisms of past practice and recommendations for the future. These questions include the compatibility of greater dialogue and openness with the broader ethos of relevant government departments. The 2000 Phillips report expressed itself on this point in very clear terms:

> Our experience over this lengthy Inquiry has led us to the firm conclusion that a policy of openness is the correct approach ... the Government must resist the temptation of attempting to appear to have all the answers in a situation of uncertainty. We believe that food scares and vaccination scares thrive on a belief that the Government is withholding information. If doubts are openly expressed and publicly explored, the public are capable of responding rationally and are more likely to accept reassurance and advice ... (1301)

This advice is clear, logical and broadly in keeping with social scientific research in this area (see, for example, Irwin and Wynne, 1996). What it now raises is first of all an empirical question about the extent to which this 'policy of openness' has successfully been put into practice in the years since 2000. Building on that point, and bearing in mind that absolute transparency is an institutional (and logical) impossibility (implying as it does both omniscience and omnipresence), how in practice has 'openness' been defined and enacted: which uncertainties are brought into public scrutiny and which are dismissed or downplayed? As the Phillips inquiry was very aware, judgements about the significance of specific uncertainties are much easier to make in retrospect than in 'real' time.

Turning to the parallel discussion of enhancing and encouraging what Tony Blair termed 'robust, engaging dialogue', how and in what circumstances do dialogue and openness accommodate with one another? Can one be enacted without the other or do they necessarily come as a package? It is certainly possible to imagine circumstances in which full and open dialogue (e.g. when exploring topics of national or institutional sensitivity) might not be assisted by a prior commitment to full disclosure. Equally, the existence of uncertainty could be used as a device for stifling robust debate (how can we discuss the future of nanotechnology when so little is known about its development and implications?) or else as a means of denying responsibility (the public has no right to complain when the risks were so clearly stated at the time). It might also be asked whether the endless listing of undeniable but remote uncertainties would serve as a distraction to public dialogue. The need for a professional judgement as to which uncertainties are worthy of discussion would seem to be unavoidable in this complex area.

Summarising these points, a number of empirical, conceptual and policy-relevant questions have emerged in the discussion so far. Three can be particularly highlighted here:

How compatible are dialogue-based approaches with conventional forms of scientific governance?

How are notions of scientific dialogue and citizenship framed within the practices of governance?

How should we conceptualise and define the current state of scientific governance?

Even these questions take us beyond what can be achieved in one chapter, but we can make at least a start by considering one study of

European practice in this area: the STAGE project (Science, Technology and Governance in Europe).

The lessons from Europe

Taking place between 2001 and 2005, and conducted by a European network of research groups, the STAGE project developed 26 case studies of scientific governance across EU eight member states. The STAGE case studies focused on three main areas: information and communication technologies, biotechnology, and the environment. Unsurprisingly given the period under study, issues of stem cell research and GM foods featured prominently. The cases particularly highlighted initiatives towards more open and dialogue-based forms of scientific governance. In marked contrast to some of the frothier rhetoric of the 'new' mode of governance, the project analysed contemporary exercises in dialogue and engagement but also suggested that more conventional approaches – governance by the market, by experts and by corporatism – remain very much dominant across Western Europe.

It is by no means a straightforward task to pull together such a complex range of experience and a variety of national contexts. The STAGE project found significant differences across the eight European countries – and even within a single country it was often impossible to identify a unitary policy style. To take the UK as one example, it is tempting to pick out for special attention a relatively few high-profile engagement initiatives and neglect the fact that these are indeed decidedly atypical of governance practice. However, the STAGE team identified several broad and strongly-overlapping features that characterise the governance of science and technology in Europe (see Hagendijk *et al.*, 2005; Hagendijk and Irwin, 2006). Whilst significant activities are taking place across Western Europe, these tend to fit within a restricted policy framework – and certainly do not suggest a dramatic shift within the practice of scientific governance. Ten particular findings from the STAGE project appear relevant in this context.

In the first place, there is a tendency across Europe to view broad public deliberation and dialogue as a one-off hurdle to be cleared at a time judged appropriate by government or scientific institutions, often quite late in the decision-making process. This sense that engagement is an activity to be initiated by policy-makers at 'the right time' has significant benefits from a planning perspective since, in keeping with conventional approaches to good governance, it allows a planned, efficient and rational approach to public engagement. However, it also suggests a very limited definition of the purpose of engagement and presents public dialogue as one discrete phase of decision-making rather than a core and abiding element the policy process.

Second, and in keeping with this observation, the STAGE studies suggest that there remains considerable 'insulation' between attempts at engagement and mainstream policy. Whilst a great deal of talk about engagement may be taking place right now, most policy processes simply continue according to their own dynamic so that, for example, conventional treatments of 'sound science' and, very importantly, science-led economic growth have remained largely unaffected. The STAGE team concluded that high-profile but atypical initiatives are generally marginal in comparison with the infrastructures dedicated to scientific/technological development.

Third, the framing of debate in Europe is typically decided by a small coterie of officials, organisations and experts of different sorts. Traditional approaches to public administration put a premium on tight organisational control, clear deadlines and careful planning (often drawing upon the advice of recognised experts and established stakeholders). Democratic engagement has a tendency however to become messy, sprawling and all-encompassing as discussion moves away from specific technically-defined topics towards, for example, issues of identity, empowerment and globalisation. This can lead to a characteristic apprehensiveness among government officials for whom expert and stakeholder consultation represents a very familiar process but the aims and outputs of public dialogue are altogether more unknown and untested. If a broader culture of engagement and external scrutiny is to be encouraged then a greater willingness to relinquish central control over the form, timing and direction of dialogue is required.

Fourth, engagement exercises are often marked by disputes over timing, organisation, and 'bias'. Certainly, open governance and enhanced dialogue are no easy solution to social contention and controversy. Whilst officials tend to see such disputes as a distraction and diversion from the 'real' questions as previously defined by debate sponsors, it might be better to accept that these are fundamental characteristics of the democratic process and as such are entirely healthy (if often uncomfortable for those under attack or caught up in the occasional political maelstrom). Once again, we can see the challenge posed by a wider engagement culture for institutions less familiar with adversarial, sprawling and contentious forms of political expression.

Fifth, and despite this evidence, it seems to be a common belief within European governance circles that dialogue and openness represent a route to consensus-building. If anything, the STAGE case studies suggest the very opposite: the more debate takes place, the more issues for discussion arise (so that this form of talk generates the demand for more talk). In analytical terms, it could also be that 'consensus' (like the concept of 'trust') needs greater deconstruction and qualification. Why, we might enquire, is consensus so prized in this area? What exactly does it signify? Meanwhile, the growing awareness that public dialogue is not necessarily an effective means of 'fixing' the 'trust problem' (as institutions still tend to perceive this) will inevitably raise further questions about the political benefits of such activities and, more broadly, of the actual aims and purposes of engagement.

Sixth, it is worth noting that deliberative governance poses challenges not only for governments, scientific organisations and industry, but also for non-governmental organisations (NGOs). NGOs often claim to speak for the public: they may even describe themselves as 'public interest' groups. Engagement exercises and more open forms of governance offer at least the potential for that claim to be undermined. Despite frequent accusations that debates have been hijacked by particular factions, public engagement can be risky for all parties.

Seventh, and put very generally, rhetoric appears to be running ahead of practice across Europe. Broad, nationwide debates are still quite exceptional. More frequent are questionnaires, focus groups and consensus conferences, but these are normally organised on an *ad hoc* basis and are not generally embedded within governance practice. Looking to the future, more incremental shifts in governance practice – the inclusion of lay members on advisory bodies, the conduct of meetings in public, the inclusion of a wider range of groups in consultation process – may have greater impact than higher-profile national debates. In making this point, it should also be stressed that rhetoric is itself a form of practice: the very fact that such language has become commonplace can be seen as a significant change in the terms of debate. Who in this era can speak against dialogue, openness, trust building and engagement?

Eighth, one important issue for the relationship between public engagement and public policy concerns the treatment of scientific evidence. In most countries under study there is a definite tendency to keep 'science' and 'the public' apart or, more precisely, to limit public engagement to what has been (somewhat problematically) defined as matters of ethics and values. If one of the great merits of engagement is the broad challenge function it offers to taken for granted assumptions and working practices, this would appear to represent a very substantial limitation on the policy process.

Ninth, the STAGE project especially identified what can be termed an 'arm's length' relationship with decision-making. Reinforcing the point above about the 'insulation' between engagement exercises and the policy-making process, one recurrent area of public and 'outsider' discussion concerned the likely impact and consequences of specific engagement exercises: was public dialogue likely to (as it was often put) 'make a difference'. Suspicion and scepticism over this point marked many European initiatives. Put in more analytical terms, the implication might be that government officials tend to see 'dialogue' as a supplement to the policy-making process rather than one of its central pillars. This finding also suggests the relative immaturity of public dialogue within science and technology policy. As we have noted, it is still approached rather cautiously by government officials, which can in turn lead to accusations that such activities are 'not being taken seriously'.

Tenth, the STAGE cases raised questions of national autonomy within the global context of innovation and the pursuit of economic competitiveness. Typically, the current practice of consultation and dialogue implies, often unquestioningly, a *national* level of decision-making. Meanwhile, it can plausibly be argued that individual nations actually exercise declining autonomy with regard to science and technology policy. In areas such as genetically-modified foods, stem cell research or nanotechnology, international agreements and market forces appear to be ever more significant. Without wishing to suggest that individual nations are necessarily powerless in the face of globalisation, the implication is that there is a profound dislocation between dialogue initiatives (which typically assume that control can be exercised within a specific nation state) and the globalising tendencies of modern science and technology.

Of course, these findings are very broad and do not do justice to the 26 individual cases studied within the STAGE project – nor is there space to discuss them here (see Hagendijk and Irwin, 2006). However, what the points above do emphasise is that deliberative initiatives tend in practice to be marginal, extremely partial and limited in scope. As such, they often represent largely isolated and insulated attempts to respond to public unrest – or else to 'fix' the perceived lack of public consensus and so pave the way for future innovation and socio-technical change. The general findings of the STAGE project certainly do little to dispel criticisms that public engagement remains a very restricted activity.

Persuasive and important though this broad conclusion might be, it certainly does not represent the end-point of analysis and reflection. In particular, such a sweeping judgement fails to address the analytical agenda outlined in this chapter. In itself, it offers neither a deeper consideration of the underlying issues nor a basis for the development of innovative approaches to public policy. It is to this 'double hurdle' that we now return.

Conclusions: reconciling public dialogue and scientific governance

Our discussion so far suggests a rather interesting relationship between public dialogue and the more routine processes of scientific governance. The 'compatibility' question lies at the very core of this relationship. On that basis, we can identify a paradox at the heart of contemporary scientific governance.

The analytical argument of this chapter is that, in turning public dialogue into an institutionally-compatible process, much of the potential challenge of that dialogue to current ways of dealing with sociotechnical change has been avoided, muted and contained. As the STAGE cases suggest, governments have tended to view public dialogue as commensurable with, but supplementary to, the 'normal' operation of scientific governance. Seen from this conventional governance perspective, dialogue-based approaches are indeed compatible and complementary. Put strongly, they can be viewed as representing no more than another option within the governance toolkit.

This has produced a situation where a study like STAGE can conclude that European (including UK) initiatives in this area have been marginal, partial and limited. Meanwhile, the rhetoric of dialogue, openness and trust-building has been enthusiastically embraced by political leaders (sometimes wearyingly so) without any apparent sense of contradiction or tension. This in turn creates a scenario where government officials can express continued disappointment that the practice of public dialogue is insufficiently relevant and useful to their operations. Meanwhile, outside critics complain that dialogue activities have been ineffective and that the outcomes are not being taken seriously. Beneath the talk of a change in the climate of scientific governance, there lies a considerable tension concerning the meaning and purpose of terms like 'public dialogue'. Put crudely, is it an incremental development of 'business as usual' or a more robust challenge to current policy processes?

Viewed in this way, we can identify how 'dialogue' has become at the same time a widely-embraced governance principle and generally marginal/partial in terms of policy impact. This finding also helps explain the phenomenon pointed to by Wynne of governmental institutions 'hitting the notes but missing the music'. In particular, Wynne argues that policy institutions have failed to include their own institutional culture within the frame of dialogue – and hence have not considered their contributory role within the generation of public mistrust and scepticism regarding technical change (Wynne, 2006). In the terms of this chapter, by converting public dialogue into a 'compatible' process, institutions have avoided the deeper challenges of 'opening up' the governance process. Paradoxically, therefore, the 'success' of public dialogue in terms of institutional incorporation (at least at the level of stated intentions) has come at the price of failure in addressing more fundamental issues of societal control and the direction of technical change. Whilst this can be presented as a fundamental criticism of scientific and policy institutions, it is also unsurprising that they should act in this way and for reasons which fit entirely with the modernistic character and history of the institutions themselves – but also their location within a wider political web of commitment to science-led growth and economic competitiveness.

A similar pattern of underlying, but unquestioned, tension can be identified with regard to the second dimension of compatibility presented above i.e. the relationship between the various strands of the new scientific governance. If we consider the linkage between public dialogue, 'trust-building' and transparency, for example, it would appear simply to have been assumed that these can indeed operate simultaneously and, moreover, that they work in support of one another. What might be viewed as an all-too-easy incorporation of such open concepts into 'business as usual' has vitiated the possibility of a closer exploration of their inter-relationship but also, more importantly, the limits to – and most appropriate form of – their applicability and operation.

Is dialogue always a good thing or are there times when it is quite simply inappropriate? What form should 'dialogue' take: a quantitative survey, a consensus conference, a telephone call? Is absolute transparency necessarily beneficial – and what exactly does transparency mean in practice? Certainly, it is not too difficult to reduce this concept to the panopticon-like absurdity of constant observation and scrutiny, and the total denial of institutional privacy and seclusion. The language of 'trust-building' is likewise problematic. As Wynne puts this: 'It is simply not possible to expect the other in a relationship to trust oneself, if one's assumed objective is to manage and control the other's response' (*ibid:* 219). The relationship between, and definition of, these concepts is indeed weakly articulated although it is also hard to avoid the conclusion that this is primarily an exercise in 'closing down' rather than 'opening up' (Stirling, 2005).

Earlier in this chapter three specific questions were raised. The first, concerning the compatibility of dialogue-based approaches with conventional forms of scientific governance, has just been addressed. What this discussion also suggests is that 'compatibility' is itself a contextually-defined term: what is compatible from the perspective of hard-pressed officials may appear less so for critics anxious to provoke a wider debate over socio-technical development. 'Compatibility' therefore begs the questions of 'with what?' and 'from whose perspective?'. This in turn can produce another rather paradoxical situation: where external critics and internal officials unite in criticising specific dialogue-based initiatives, albeit from very different premises. Viewed in this way, the risk becomes apparent that public dialogue, despite its general appeal in the abstract, will end up pleasing nobody when weakly translated into institutional practice.

Moving to the second question, how then are notions of dialogue and citizenship being framed in this context? Above all, this chapter has identified the rather impoverished frameworks now being employed and developed within scientific governance. Turning the compatibility question around, it is of course true that wider dialogue and debate over such matters as nanotechnology, biotechnology and new energy futures might raise questions that are too large for current institutional and policy-making structures to deal with.

Although administrative expediency would suggest cutting these discussions down to what can be practically dealt with, this is not necessarily the best way forward. It may instead be time to step outside the accepted routes for the development of science and technology policy and consider what alternative frameworks might exist. This is especially true when those accepted routes might be part of the very problem with which public dialogue must deal.

What does all this mean for our conceptualisation of the current state of scientific governance? The last ten years of British experience suggest a pattern of both change and continuity. Without wishing to deny the very considerable changes that have taken place, it may well be that continuity has been the stronger feature – especially in terms of the underlying institutional commitment to building 'trust and confidence in the way that science can demonstrate new opportunities, and offer new solutions' (Blair, 2002). This certainly offers a very loaded and instrumentalist framework for attempts at dialogue and engagement.

However, the new rhetoric of dialogue has also created new opportunities and possibilities. It is here that the 'double hurdle' becomes especially significant. Lack of critical reflection and debate about the underlying character of public dialogue has gone hand in hand with an incremental and restricted policy response to such issues. Of course, other factors are at work also – not least issues of political economy and the broad governmental commitment to science-led innovation. Nevertheless, the challenge of scientific governance to both academic scholarship and institutional practice cannot be denied. This chapter represents simply one attempt at addressing this considerable challenge.

References

Tony Blair, Science Matters. April 10, 2002. Available from: www.number-10.gov.uk/output/Page 1715.asp [accessed 17 November 2006].

R. Brickman, S. Jasanoff and T. Ilgen, *Controlling Chemicals: the politics of regulation in Europe and the United States* (Ithaca: Cornell University Press, 1985).

Council for Science and Technology, 'Policy through dialogue: informing policies based on science and technology'. A report from the Council for Science and Technology, March 2005: www.cst.gov.uk/cst/reports/#8 [accessed 17 November 2006].

H. Collins and T. Pinch, *The Golem* (Cambridge: Cambridge University Press, 1993).

Department of Trade and Industry, *Excellence and Opportunity: a science and innovation policy for the 21st century* (London: The Stationery Office, 2000).

European Commission, *Science and Society: Action Plan* (Luxembourg: Commission of the European Communities, 2002).

R. Grove-White, 'New Wine, Old Bottles? Personal reflections on the new biotechnology commissions', *The Political Quarterly*, 72(4) (2001) 466–472.

R. Hagendijk, P. Healey, M. Horst and A. Irwin, 'Science, Technology and Governance in Europe: challenges of public engagement', *STAGE Final Report*, February 2005.

R. Hagendijk and A. Irwin, 'Public deliberation and governance: engaging with science and technology in contemporary Europe', *Minerva*, 44 (2006) 167–184.

House of Lords Select Committee on Science and Technology, *Science and Society* (London: The Stationery Office, 2000).

A. Irwin, *Risk and the Control of Technology* (Manchester: Manchester University Press, 1985).

A. Irwin, 'The politics of talk: coming to terms with the "new" scientific governance', *Social Studies of Science*, Vol. 36(2) (2006) 299–320.

A. Irwin and B. Wynne, *Misunderstanding Science?* (Cambridge: Cambridge University Press, 1996).

Office of Science and Technology, *Guidelines 2000: scientific advice and policy making*, DTI, London: www.dti.gov/uk/science/page15432.html [accessed 17 November 2006].

A.M. Pettigrew, 'The character and significance of management research on the public services', *The Academy of Management Journal*, 48(6) (2005) 973–977.

Lord Phillips, J. Bridgeman and M. Ferguson-Smith, *The BSE Inquiry: The Report* (London: The Stationery Office, 2000).

Royal Commission on Environmental Pollution, *Setting Environmental Standards*. 21st report (London: The Stationery Office, October, 1998).

Royal Society/Royal Academy of Engineering, *Nanoscience and Nanotechnologies: opportunities and uncertainties*. RS policy document 19/04. July 2004 (London: Royal Society).

A. Stirling, 'Opening up or closing down? Analysis, participation and power in the social appraisal of technology', in Leach, M., Scoones, I. and Wynne, B. (eds) *Science and Citizens: globalization and the challenge of engagement* (London and New York: Zed Books, 2005) 218–231.

B. Wynne, 'Public engagement as a means of restoring public trust in science – hitting the notes, but missing the music?', *Community Genetics*, 9(3) (2006) 211–220.

3
On the Signature of New Technologies: Materiality, Sociality and Practical Reasoning

Tom Horlick-Jones

Introduction

In this chapter I will examine critically some of the underlying socio-logical ideas present in recent debates about the social acceptability of new technologies. I focus on the notion of constructionism: a perspective and analytical approach that recognises, and seeks to explicate, the ways in which the categories of human discourse are socially negoti-ated and selected (see e.g. Hannigan, 1995). I will argue that whilst the use of constructionist ideas has enriched such debates, and moved them away from a narrow technocratic reductionism, they have done so at the risk of losing track of the specific features of technological artefacts. In seeking to include human sensibilities in the analysis, a preference has been given to sociological theories of reality at the expense of engaging with what I will call the *signature* of the techno-logy: the specific ways in which it is articulated in practical reasoning and discourse within real-world settings.

Social constructionist accounts – and in due course I will discuss some of their diversity – share the characteristic that they recognise a slippage between what one might call 'conditions' and 'claims'. In other words entities that may seem fixed and immutable – as diverse as ecology, illness and the constituent particles of matter (some of many examples from the literature cited in Hacking, 1999) – may come to be regarded differently, yet equally 'solid' in nature, when, in different cir-cumstances, they are viewed through different socially-organised 'ways of seeing'.

There is a clear resonance here with many matters of public debate which turn on the assessment of risk. Experts give advice on the basis of technical assessments of conditions, in the light of which the likes and dislikes of lay publics may appear 'irrational'. In their path-breaking work on the cultural basis of risk perception, Douglas and Wildavsky (1982) considered why, in the United States, many techno-logical issues – like the management of nuclear waste and regulation of chemicals – were associated with controversies and unresolved conflicts. They famously posed the question (*ibid*: 1): 'Are dangers really increasing or are we more afraid?'.

A little over a decade later, the features that had characterised many American technological controversies at that time had become more common in Europe. In a collection of papers associated with a European Commission-sponsored conference on scientific expertise and public policy, a colleague and I argued that that some kind of turning point had been reached in the capacity of certain risk-related issues to mobilise European lay publics (Horlick-Jones and De Marchi, 1995; see also Horlick-Jones, 2004). At the time of writing, the dispute over the decommissioning of the Brent Spar oil platform in the North Sea had become a prominent feature on television screens and in news-paper headlines. The controversy had also prompted a mass consumer boycott and other forms of direct action in a number of European countries (Löfstedt and Renn, 1997). A polarisation within policy and lobbying circles was taking place. For some, US-style single-issue pres-sure politics had crossed the Atlantic, promising an anarchic future for the control of technology. For others, the mass action represented a spectacular assertion of popular democracy; a shot across the bows of multinational corporations and government bureaucracies.

Seen with the benefit of hindsight, the Brent Spar controversy might indeed be regarded as something of a watershed. The extent to which it prompted shifts in market behaviour rang alarm bells in government offices and corporate boardrooms alike. In the years that followed a similar pattern of media headlines, contested expertise, and an insta-bility in consumer behaviour often came to characterise the form of controversies associated with new technologies.

To the extent that new animal feeding practices may be regarded as a technological innovation, the Bovine Spongiform Encephalopathy (BSE) outbreak in British beef was perhaps the most spectacular such crisis. Of course, in this case, it transpired that the consequences were only too concrete, and tragic. However, in the case of the claims regarding the dangers of the Measles, Mumps and Rubella (MMR) vac-cination (Bellaby, 2003), the health hazards of mobile telephones (Burgess, 2004) and the health and environmental impacts of genet-ically modified crops and food (Horlick-Jones *et al.*, 2007a) the evidence of possible harm has not been so decisive. Nevertheless these latter controversies have had the potential to themselves generate harm: for instance, a low take-up of the MMR vaccination in the UK could still lead to a damaging measles epidemic. More generally, echoes of the sort of fears voiced in the USA in the 1980s and 90s about an 'over-anxiety' about the dangers of technology (Huber, 1991; Glassner, 1999) have now been heard in Europe. Such 'irrational' fears, it is claimed, will bring about business failures, a lack of economic competitiveness, lost opportunities for innovation, and other pathologies (Furedi, 1997; Durodié, 2003; Burgess, 2004).

These changes in the nature of technological controversies have been accompanied by significant developments in public policy discourse within many democratic countries. In particular, the need to 'engage' with lay publics – to be responsive to their views, and to involve them in decision-making – is now regarded as an important component of effective governance. In practical terms, this is seen as addressing a number of perceived crises faced by contemporary governments, arising from insufficiencies of knowledge, trust and legitimacy (CEC, 2001; OECD, 2001). In Britain, the public inquiry into BSE (Phillips Report, 2000) called for radical new ways of thinking about, and managing risk. Proposed measures included a far higher degree of openness in government as a pre-condition for building trust and credibility in official policy and practice. A few months earlier, the House of Lords' report on *Science and Society* (2000) had signalled a decisive move away from a 'deficit' model of public (mis)understanding of science – which stresses public ignorance of technical facts – towards an advocacy for engagement with lay publics, their views, and their values.

Interestingly, the House of Lords report (*op cit*) adopted a very different position with respect to social science than an earlier, very influential, report on the public understanding of science, produced 15 years before by another pillar of the British establishment: the Royal Society (The *Bodmer Report*; 1985). For Bodmer and his colleagues, scientific knowledge was seen as essentially neutral, unproblematic stuff, concerned with truth about the physical world. It was provided by experts, and effective public understanding was to be achieved by effective targeting and presentation, so as to 'fill up the minds' of lay audiences. According to this model, the appropriate role of social scientists in promoting public understanding of science amounted to that of glorified marketing personnel, surveying attitudes and assessing the scope for 'selling' this particular commodity (discussed in Horlick-Jones, 1998a). In contrast, the position adopted by the House of Lords was strongly influenced by part of the literature that draws heavily on a number of strands of constructionist thinking about science, technology and risk (e.g. Irwin and Wynne, 1996).

Just two years later, in an unprecedented development, a process of government-sponsored public debate (branded *GM Nation?*) about the possible cultivation of genetically-modified crops in the UK began. Despite suffering from a number of imperfections, the debate was successful in generating widespread interest and considered discussion about complex matters of science and policy amongst relatively large numbers of the lay public. This would have been unthinkable in policy circles just a decade before (see Horlick-Jones *et al.*, 2007a). The decision to hold the debate may be understood partly in terms of a response to BSE and other high-profile government and corporate failures that had occurred over the previous decade. More generally, it took place in a context in which the British state was attempting to reconcile tensions within the nexus of scientific and technological developments, social values and expectations, and market pressures (Walls *et al.*, 2005). It also seems likely to reflect a pre-occupation within British government circles about the need to 'regain' a perceived loss of trust in public institutions (Löfstedt and Horlick-Jones, 1999; Walls *et al.*, 2004). Importantly, these changes in governance had created a context that allowed constructionist thinking to be incorporated into the highest levels of policy-making.

In the following section, I examine some of the kinds of constructionist analysis to be found in the literature, and I introduce an analytical typology that seeks to find some form in this diversity. I then use this typology to discuss two contrasting constructionist approaches to understanding the social acceptability of technologies. The fourth section discusses questions of constraint and opportunity that arise from the material nature of technologies. It also introduces the notion of what I term a risk issue's 'signature': a measure of its capacity to engender certain patterns of understanding. The fifth section pursues this notion of signature in more detail by focusing on how lay understandings of technology are 'talked into existence' in practical discourse. Finally, I bring the chapter to a end with some concluding remarks.

The labyrinth of constructionist analyses

As a number of writers (e.g. Hacking, 1999; Lynch, 2001; Velody and Williams, 1998) have observed, analyses that go by the name of constructionist, whilst possibly having certain shared characteristics, display enormous variety in terms of perspective, method and objectives. They variously embody notions of framing, labelling, defining, categorising, selecting and sense-making, and arguably it is sometimes difficult to discern whether the character of a given analysis is primarily concerned with ontological, epistemological or methodological matters. I begin this brief, and somewhat breathless, discussion of constructionism by considering perhaps the most influential such account of technology and risk: that of the Risk Society.

In recent years an increasingly conventional view within sociology has been the argument that processes of modernisation have engendered an increased reflexivity within social agents, leading to individualisation, detraditionalisation, and a chronic sensitivity towards risk-related issues. According to this theory of *reflexive modernisation*, slightly different versions of which are together often known as the Risk Society thesis (Beck, 1992; Giddens, 1991), this state of affairs leads to an erosion of trust in sources of expertise, which are experienced in the form of abstract, depersonalised, and all-too-fallible 'expert systems'. In this way the Risk Society theory seems to offer a way of understanding a shift in lay sensibilities towards many health and environmental issues that has taken place in recent years. It also suggests that an anxiety about innovation may serve to create resistance to certain new technologies.

Despite its considerable influence within scholarly circles, first in Europe and then internationally, and to some extent within policy circles, the Risk Society thesis has received severe criticism. For example, critics point to its alleged German parochialism (Dingwall, 1999), its simplistic and one-dimensional account of subjectivity (Rose, 1996) and its inability to account for the diversity of risk-related practices in the real world (Horlick-Jones, 2005a). However it is the critique advanced by Barnes (1995) that I wish to single out for special attention here. This addresses Beck's version of the theory, in which the 'motor' of change is seen as the pervasive and pathological by-products of industrial production. For Beck, the ever-increasing threats to the survival of the planet, posed by risks that endanger rich and poor alike, lead to demands for risk reduction that cannot be satisfied, and so a delegitimating spiral is produced.

In a memorable phrase, Beck (*op cit*: 55) argues that:

It is not clear whether it is the risks that have intensified, or our views of them. Both sides converge, condition each other, strengthen each other, and because risks are risks in *knowledge*, perceptions of risk and risk are not different things, but one and the same.

The slippage between 'conditions' and 'claims' that I have identified as a characteristic of constructionist analyses is evident here. However, there are many versions of constructionism, and Barnes goes on to diagnose Beck's version as 'idealist'. In other words, he is suggesting that it draws upon a philosophical tradition that holds the world is a thing of the mind, and that there exists no reality independent of socially-organised representations. Barnes (*op cit*: 110) ridicules this perspective in devastating fashion:

> ... to take literally the equivalence of perception and actuality asserted by Beck suggests the possibility of some radically new approaches to improving the environment of which he would be unlikely to approve. If getting the lead out of drinking water is expensive, why not stick some valium in the water instead? If the educated middle classes are increasingly suffering from acute anxiety about ecological uncertainties, how about free psychotherapy for everyone entering higher education?

Barnes goes on to suggest that Beck's programme might not be best served by such a version of constructionism. Here he is referring to the politics of Beck's work, and what he regards as its motivating commitment to environmentalism. At stake here is the underlying reality of the issues whose constructed nature is being considered. But this is not just a problem for environmentalists. Critics from a range of political and conceptual perspectives have argued that constructionist analyses seem to call into question whether 'real' problems 'really' exist beyond the politics of competing stakeholder perspectives (e.g. Anderson and Mullen, 1998; Norris, 1995; Sayer, 2000).

Historically, the term 'social construction' seems to have first appeared in Berger and Luckman's (1966) well-known reading of Schutzian phenomenology. However the link between this work and subsequent constructionist analyses is rarely clear. Arguably more important, in terms of their influence on thinking about technological risk, has been the 'social problems' literature which originated with the work of Spector and Kitsuse (1987); and the sociology of scientific knowledge (SSK: Lynch, 1993; Pickering, 1992) and social construction of technology (SCOT: Bijker and Law, 1992) literatures. Whilst the former has been concerned with the dynamics of how social problems come to be seen as such; the latter has investigated the ways in which social factors shape the form of scientific knowledge, and of technologies. The SSK and SCOT literature is sometimes described as 'constructivist', however for the purposes of this chapter, I will use the word constructionist in a generic way to embody both social problems and SSK/SCOT forms of analysis.

Constructionist views of risk recognise that the identification and assessment of risk is a social activity and, as such, is concerned with the production of meaning and a shared understanding of reality (Hilgartner, 1992; Horlick-Jones, 1998b). The role of the social in the construction of meaning needs to be invoked in order to appreciate how different societies, and indeed subcultures, sometimes have radically different beliefs and sense of what is real and true. This multiplicity of meaning lies at the heart of why a given risk is sometimes perceived by different social groups as posing a very different degree of threat (or opportunity). This approach can lead to a number of different conclusions about the nature of risk. Within the literature, what has been described as a *strong constructionist* analysis corresponds to the argument that nothing is a risk in itself, and that what is understood as such is a product of historically, socially and politically contingent perspectives. A *weak constructionist* position would recognise risk to be an objective hazard, threat or danger, however one that is inevitably mediated through social and cultural processes (see Fox, 1998; Lupton, 1999; Renn, 1992).

In conceptual terms, these debates raise some profound questions concerned with the relationship between some entity's ontological nature and the categories that are used to describe it. To what extent does that entity's innate nature demand its description takes a certain form; and, conversely, to what extent do ideological and value 'spin' impose a given way of seeing? The dynamic in question here is therefore a *tension between materiality and sociality*, or, more fundamentally, between constraint and opportunity. I will return to this theme.

An important practical characteristic of constructionist analyses is their capacity to be used normatively, or as Barnes (1995: 106) puts it, 'as a weapon in the defence of particular perceptions and policies'. A phrase that often occurs in these writings is 'things could have been otherwise'; the implication being that the apparently essential features of an issue in question are rather less solid and fixed than commonly believed. This recognition has led to much work that has sought to deconstruct notions of 'natural order' by demonstrating their social and ideological roots. Whilst questioning possibly bogus claims of essentialism seems an important task for sociology, there is also a clear danger here of romantic readings that reflect the certainties of underlying value commitments on the part of the analysts (see e.g. Hacking, 1999; Horlick-Jones and Sime, 2004).

Perhaps the most pungent criticism of this tendency has been advanced by Woolgar and Pauluch (1985), who have argued that constructionist analyses are typically accomplished by the problematising of certain phenomena whilst leaving other unproblematic. This, they argue, amounts to a selective relativism which privileges certain objectivist assumptions; a process they term 'ontological gerrymandering'. In response to the debate prompted by this intervention, Best (1989) proposed a three-fold categorisation of types of constructionist analysis. The first, which he calls *strict* constructionism, avoids making the sort of objectivist assumptions to which Woolgar and Pawluch draw attention. Such analyses are relativist in the sense that they do not seek to adjudicate between competing accounts of the issues in question. A second category, *debunking*, is described by Best as a 'crude' method, in that it seeks to demonstrate false claims by drawing upon the analyst's 'known' objective reality. Finally, Best identifies *contextual* constructionism, in which the analyst gathers other evidence which may have a bearing on the status of the claims-making being scrutinised.

As we will see, this categorisation is of an ideal-type variety, and in practice constructionist analyses often take a complex, hybrid, form. Moreover, their status may be contested. Take the case of constructionist writings about environmental issues. Here, as noted above, some realist critics point out that such work casts doubt on the very existence of such issues. However Burningham and Cooper (1999) suggest that that most of these constructionist analyses are, to use Best's term, contextual in character. They go on to argue that in choosing 'extreme' (or in Best's terminology, 'strict') constructionist analyses for attack, critics are, in effect, attacking a 'straw man'. Despite such considerations, I suggest that Best's scheme provides a very useful analytical perspective and vocabulary and, in the following section, I shall attempt to apply it to examine the work of two prominent contributions to sociological work on the social acceptability of technology.

Two approaches to understanding the social (un)acceptability of technologies

As mentioned above, recent years have seen an important shift in the extent to which sociological ideas about technological risk have had a role in policy discourse. Until recently, public debate about such matters took a form that was largely dominated by technical knowledge and instrumental rationality; as in the case of public inquiries into the building of nuclear power stations (O'Riordan *et al.*, 1988; Wynne, 1982). In 1992, the refusal by the Royal Society of London to endorse the report of its study group on risk, which included two chapters by social scientists, was nicely captured in Orwellian terms as a case of 'four chapters good, two chapters bad' (Hood and Jones, 1996: xi).

Of course, sociologists do not speak with one voice, and here I consider briefly the work of two important, and contrasting, contributions to the literature. The first is by Brian Wynne, who has been a well-established, and influential, commentator over a period in excess of 20 years. He was appointed a specialist adviser to the House of Lords committee that produced the *Science and Society* report (*op cit*). The second contributor, Adam Burgess, has quickly established his reputation as a critic of over-zealous regulation since the publication of his book (2004) on mobile telephones and what he calls the 'culture of precaution'.

For Wynne (e.g. 2001, see also Irwin and Wynne, 1996; Lash *et al.*, 1996), disagreements over risk issues between technical experts and lay people arise because these different groups have different ways of *framing* the issues. Rather than being irrational or unscientific, he argues that lay groups typically include a wider range of considerations in their reasoning processes; issues that are of relevance to the tasks of everyday life. Such considerations include a rich sphere of mundane experience, allowing judgements to be made, for example, on the extent to which organisations responsible for managing the risks in question may be trusted.

In what is perhaps his best-known study, Wynne examined the social relations between Cumbrian sheep farmers and government scientists after the Chernobyl nuclear accident, and the circumstances in which inadequate official guidance about radioactive contamination was issued. The contamination of sheep threatened to rob the farmers of their livelihoods, however, according to Wynne, additional threats were posed by the government action itself (quoted in Horlick-Jones, 1998a: 323):

> Naturally the farmers felt their whole identity was under threat from outside interventions based upon what they saw as ignorant but arrogant experts who did not recognise ... their specialized hill-farming expertise.

It is not possible to do complete justice to the complexity of Wynne's thought within the confines of this chapter. However a number of important features can be identified. His approach embodies a number of themes: social framing; the limits of, and politics of, expertise; and elements of the sociology of scientific knowledge. Importantly, it contains the idea that alienation can produce a sort of grounded and reflexive perspective that is not available to the powerful (see discussion in Horlick-Jones, 1998a; Mulkay, 1997). There is also the use of social constructionist ideas about identity (see e.g. Shotter and Gergen, 1989). Importantly, despite drawing on a literature that views identity as 'intrinsically incomplete' (Wynne, quoted in Horlick-Jones, 1998a: 323), his work is rooted in the idea of an essential humanity – or as he puts it (Wynne, 1996: 381): 'human universals beyond the scientistic imagination'. In this way, 'public concerns' about new technologies like genetic modification are seen as arising from the alienation created by powerful, technocratic framings of risk issues and patronising views towards lay publics (Wynne, 2001).

In contrast, Burgess' work (2004) on the fears associated with mobile telephones is a sophisticated but more orthodox form of 'social problems'-type constructionist analysis. He has collected a wealth of evidence for the case that contemporary anxieties about risk are driven by media coverage 'fuelled from above' (*ibid*: 13) by institutional and political interests. Such bodies need to maintain their legitimacy, and so seek to avoid the risk of being seen to fail to protect the lay public. In this sense, his view chimes with recent work which has identified the management of such 'meta-risks' as a central concern of contemporary governance (Horlick-Jones, 2005a; Rothstein, 2006; see also Rose, 1999). In terms of his underlying view of humanity, there is a suggestion of essentialism here, in that human beings are regarded as 'extraordinarily robust and capable of adaptation' (*ibid*: 281), with the pursuit of 'phantom' risks threatening to compromise and diminish what he terms 'our autonomy, intelligence, and capacity for change and enlightenment' (*ibid*).

In the space available I have only been able to sketch these two contributions to the sociology of resistance to technology. Both are clearly constructionist in nature in that they explore the underlying mechanisms that create a slippage between 'conditions' and 'claims'. To what extent, though, can they be categorised according to Best's typology? Neither analysis takes a strict constructionist form, in that both adjudicate between accounts. Neither is crudely debunking, although both introduce external sources of essentialism in the form of expectations about human nature. There is also a politics to their positions; with Wynne committed to what might be described as the 'democratisation of science and technology' (Irwin and Wynne, 1996; see also discussion in Horlick-Jones, 1998a), and Burgess sharply critical of precautionary-style regulation. Both gather contextual information, with Burgess arguably appealing to a wider range of sources and perspectives.

Importantly, these analyses have further features in common. They are both realist; Burgess clearly so, and Wynne in a way reminiscent of Douglas' work. She argues (e.g. in Douglas, 1990) that certain risks are selected for special attention by groups primarily because of the threat or support they present to preferred *ways of life*, rather than any inherent objective measure of hazard that they pose. However she is at pains to make clear that (*ibid*: 8): 'The dangers are only too horribly real ....This argument is not about the reality of the dangers, but about how they are *politicized*' [stress added].

Indeed, Wynne and colleagues (Lash *et al.*, 1996: 10) are explicit in stating that his focus on the social relations of technology means that: 'whether the risks "actually" physically exist is irrelevant to this dynamic'. In this way, such theories chime with the recognition that risk issues lend themselves to symbolic associations and the generation of a rich micro-politics (discussed in detail in Horlick-Jones, 2005a; 2005b).

Crucially, both Wynne and Burgess make strong assumptions about the form of underlying lay rationalities and the degree of knowledge possessed by lay publics. For Wynne, lay publics are alienated by the form of expert discourses of risk; whilst for Burgess, over-zealous regulation itself engenders concern. Both assume that lay publics are aware of official risk policies, at least to the extent that this information prompts expressions of resistance or anxiety. Arguably, both authors' conception of lay rationality possesses a certain *monolithic* character; by which I mean a way of reasoning that is not dependent upon the specific features of the technologies in question, or of the contexts in which these technologies are apprehended.[1]

The central features of these theories resonate with the focus of a radical critique of constructionist accounts of technology produced by scholars like Button (1993) and Hutchby (2001). According to Button (*op cit*: 9), in such accounts 'the technology disappears from view'. It does so because of their 'insistence' (*ibid*): 'on subduing the articulation of technology in technical and mundane discourse to a sociological theory of reality'. Button's critique draws heavily on ethnomethodology; which is, of course, a form of sociology that is concerned with practical reasoning in everyday situations, and the ways in which practical actions are collectively accomplished and made intelligible (Garfinkel, 1967). The form of the critique reflects the contrast that ethnomethodology draws between conventional sociological accounts (which are fundamentally 'about' sociology) and its own attempt to *respecify* sociology by capturing social phenomena 'from within' (Button, 1991; Lynch, 1993).

Importantly, this critique of constructionist accounts of technology points to the need for empirical studies of the fine detail of lay practical reasoning about technology in specific everyday situations, and how the specific features of technologies are embedded in the form and logics of such talk. In this way, it is possible to capture the *haeccity* or 'just this-ness' (Garfinkel, 1967; Lynch, 1993) of the phenomenon in question. Later in the chapter I will examine some empirical examples of such practical reasoning. First, in the following section, I will consider the question of materiality, and the extent to which constructionist accounts embody the constraints and opportunities produced by a technology's material nature.

The limits on malleability: materiality as an opportunity structure

Earlier in this chapter I noted that risk issues entail a tension between materiality and sociality, or, more generally, between constraint and opportunity. It is pertinent to note that many sociologists have been critical of technocratic reductionism, but seem content to be involved in the methodological 'reverse side of the coin': namely social reductionism. This tendency is nicely analysed in Strong's (1979) celebrated essay on 'sociological imperialism'. As Strong and Dingwall (1989: 53) subsequently noted, 'sociologists have been too quick to disregard the material limits to human action' (see also Sayer, 2004). Much diverse work by social scientists has attempted to capture the specific features of technologies, and how their characteristics have the capacity to create constraint and provide opportunity (e.g. Kaldor, 1982; Molotch, 2003; Perrow, 1984; Collingridge, 1992). Constructionist analyses have shown how human appreciation of the material world entails the use of categories that are made available, and shaped, by social institutions through which everyday life is lived. However there is a need to recognise that the material world is not infinitely malleable.

This sense of structure is captured by the notion of *affordance*, which was introduced by the psychologist J.J. Gibson (see also Anderson and Sharrock, 1993; Sime, 1999; Hutchby, 2001), as a measure of the capacity of the material world to provide opportunities for action. To use

Gibson's term, technologies *afford* certain possibilities for their use. Those uses reflect not only the material character of the technology, but also the inclinations and circumstances of those who are making sense and use of the technology. In its original psychological terms, affordance (Gibson, 1979: 129): '... is neither an objective not a subjective property; or it is both if you like'.

Sociologists who have found the idea of affordance useful have adapted this definition somewhat – recognising that the sense-making dynamic draws upon socially-available categories – but the essential notion remains the same. One might think of a technology's affordance as placing limits on the social and cultural imagination.

One way of possibly approaching the affordance of technologies in empirical terms emerged from recent work I conducted with some colleagues (Petts *et al.*, 2001). We were examining evidence in lay discourse about various risk issues, (mostly associated with technologies), for the suggestion that the media has a role in enhancing the degree of perceived threat posed by these issues. We recognised the methodological dangers inherent in simply taking at face value respondents' accounts of the influence of media sources on their views. So we chose to analyse the corpus of focus group-generated data we had collected by regarding participant accounts as *topics* in themselves, rather than as *resources* that provided unproblematic information (cf. Garfinkel, 1967). We developed an analysis of the structure of these accounts by examining the ways in which speakers warranted the claims they were making. Were they appealing to knowledge derived from everyday experiences, the experiences of families friends and colleagues, or media sources? This analysis led to the finding that different risk issues engender talk that is characterised by different distributions of references to these three sources of knowledge.

Some of our findings are shown in Figure 3.1. Here the account structures of three of the risk issues we investigated – train accidents, possible adverse impacts of genetically modified (GM) food, and industrial air pollution – are compared. The vertical axis measures the percentage of accounts that were warranted according to the three sources of knowledge mentioned. We found that talk about train accidents was dominated by references to mediated knowledge. This is not surprising as few people have personal experience of these accidents, or indeed know anyone with such experience. In contrast, talk about air pollution was strongly grounded in terms of direct experience, and included many anecdotes which invoked evidence of this hazard. In the case of GM food, participants struggled to ground the frequent allusions to

Figure 3.1   Appeals to sources of knowledge in accounts of three risk issues

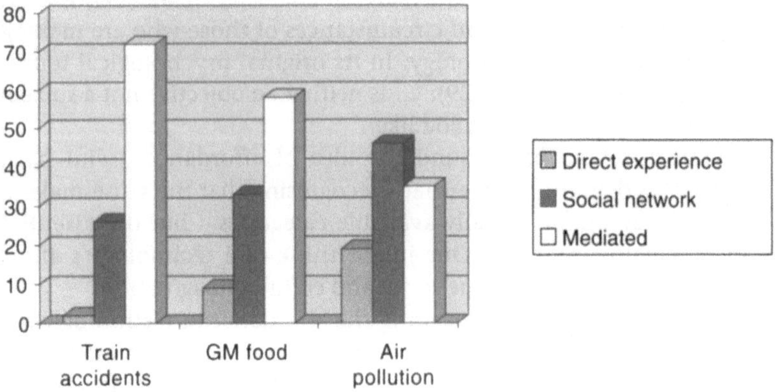

Here the percentage of accounts warranted by these sources are shown for each of the three issues (adapted from Petts, Horlick-Jones and Murdock, 2001).

mediated knowledge in the talk in terms of everyday, experience-based understandings.

Our examination of these account structures led us to speculate (*ibid*) that different risk issues have different capacities to engender specific patterns of understanding and response. We chose the term *signature* to denote this capacity. A comparative analysis of talk about different risk issues provided further insights into the nature of the factors that generate signatures. Four factors had a particularly strong role in the ways in which the groups made specific sense of risk issues: the specificity of possible ill effects; the extent to which concern for others was engendered; whether some degree of secrecy or cover-up was associated with the issue; and whether the issues entailed ethical concerns. Here one can see a resonance with the subjective factors identified in the now-classic psychometric investigations of risk perception (Slovic, 2000). In this way, the notion of a signature may provide a way of relating such psychological findings to analyses which are located at the level of social interaction.

We also found (*ibid*) that the perceived trustworthiness of the institutional framework responsible for managing a given risk issue had an important bearing on how that issue was regarded. This finding from the work therefore provides some support for Wynne's and Burgess' focus on the 'institutional body language' of government and regulatory organisations. However, it is important to note that this dimension alone does not capture the full richness of a given risk issue's signature. Rather, the signature seems to embody something of how such issues are articulated in terms of practical reasoning within everyday discourse, so reflecting both their materiality and the sociality of the circumstances in which they are apprehended.

Technologies in mundane discourse

In my discussion of Wynne's and Burgess' work, I noted that both seem to assume what I termed a monolithic form of rationality on the part of lay publics. They also seem to assume that lay audiences possess a very high degree of knowledge about the character and actions of official policy-making and regulatory bodies. These assumptions are brought into question by recent work on risk, practical reasoning and social interaction (e.g. Bloor *et al.*, 2006; Boden, 2000; Candlin and Candlin, 2002; Horlick-Jones, 2005a; 2005b; Horlick-Jones *et al.*, 2007b; Maynard, 2003; Timotijevic and Barnett, 2006; Walls *et al.*, 2004). This literature points to the existence of situationally-specific, and emergent logics entailed in shaping risk reasoning across a range of organisational and social contexts. Other recent work (Horlick-Jones *et al.*, 2007b; Walls *et al.*, 2004) suggests that these diverse modes of reasoning also reflect differing degrees of understanding of the issues in question.

In this section I illustrate some of these features of practical reasoning about risk-related issues. In so doing, I seek to explicate the nature of given technology's signature in terms of the practical accomplishment of sense-making, as it occurs in the moment-by-moment unfolding of social interaction. I draw upon a detailed analysis of lay reasoning about genetically modified (GM) crops and food which has appeared elsewhere (Horlick-Jones *et al.*, 2007b; see also Horlick-Jones *et al.*, 2007c). The data I use to illustrate these features was generated during the course of the recent public debate in Britain about the possible commercialisation of GM crops (Horlick-Jones *et al.*, 2007a). Specifically, it comprises talk produced by a series of ten reconvened discussion groups, which formed a component part of the debate process. These occasions took participants through a learning process in which they were sensitised to some issues concerning genetically modified crops and food, exposed to various sources of information including media accounts, and invited to discuss their emerging understandings. The groups therefore provided an invaluable setting in order to study a developing process of learning and sense-making about a given technology.

The first sequence of talk I wish to consider is reproduced at Figure 3.2. The interlocutors are discussing whether the consumption of GM food presents the possibility of any adverse health effects. In lines 1–4, M1 makes the point that such effects might not be detectable for an extended period of time. F1's turn, which starts at line 5, develops this theme by analogy with 'mad cow disease' (BSE), the long-term impact of cigarette use, and possible adverse side effects of hormone replacement therapy. At this point, in line 10, M4 adds mobile phones to the list being considered by the group. He uses the term 'another one isn't it', suggesting an assumption that possible health impact issues associated with this technology will be understood by the other participants, and this technology will be seen by them as an appropriate addition to the category they are building. At line 11, F1 reinforces this sense of shared understanding by attempting to formulate the category as 'issues you think if you knew straight away would you risk it ...'.

At line 13, F2 questions whether people really wish to know about these dangers. F1 responds by saying she does wish to know, but notes that the BSE threat has not prevented her children from eating burgers. This comment may be heard as indicating that despite the perceived threat she has not felt it was sufficiently threatening to prevent, or dissuade, her children from eating burgers. In making this

Figure 3.2   An illustration of the dynamics of talk about GM issues

NBD/4/12-13 (M$_i$ = male participants; F$_i$ = female participants)		
1	M1	I think the one worry that I have is are there any side effects I think we just don't
2		know and we might not know for twenty years I don't know but until somebody
3		says they've had a funny turn because they've been eating GM food all their lives
4		I'm not sure but yeah
5	F1	It's a bit like that mad cow disease isn't it you know well it's like anything would
6		we have handed out cigarettes to the troops in the war if you'd known they were
7		going to cause lung cancer forty odd years down the line you wouldn't have done
8		it you know HRT for women going through the menopause makes them feel
9		absolutely wonderful now but what's the long term effects of that ((gap))
10	M4	mobile phones is another one isn't it
11	F1	yeah there's so many that ... issues that you think if you knew straight away
12		would you risk it and then
13	F2	do you want to know
14	F1	but then yes do you want to know I mean with mad cows disease it hasn't
15		stopped my kids eating burgers and or the egg thing you know when Edwina
16		Curry did her sort of thing with that
17	F3	or is everything good in moderation
18	F1	yeah as long as you don't overdose on things but its knowing it's getting more
19		facts to be able to make a decision

Adapted from Horlick-Jones *et al.*, 2007a.

statement, she risks presenting herself as acting irresponsibly towards her children. She immediately mentions a controversy that took place in the UK about the safety of eating eggs, in which a prominent politician who advised caution was severely criticised for over-reaction. In this way, she illustrates the kind of moral dilemma with which she has struggled.

The intervention at line 17 illustrates one means by which the process of making sense of technologies and risk issues is collectively accomplished. Here F3 uses the term 'everything good in moderation'. This might be regarded as something of a cliché. However it is also an example of what has been termed a *lay logic*, or *lay logical device* (Horlick-Jones, 2005b; Horlick-Jones *et al.*, 2007b; Petts *et al.*, 2001; see also Myers and Macnaghten, 1998). Such devices, which take the form of generally accepted arguments, seem to be used in everyday talk as shared interpretative resources that assist making sense of evidence about some issue.

At lines 18–19, F1 responds to F3's intervention by noting that what constitutes 'moderation' may be uncertain. This exchange between F1 and F3 may be heard as seeking to reducing the possible culpability implied by F1's initial comment about burgers, by re-establishing a sense of uncertainty over whether F1 and, by implication, other parents present, has behaved responsibly.

We rejoin this group conversation a few moments later in Figure 3.3. The topic has moved to mobile telephones. It should be noted that in such relatively unstructured talk about risk issues, it is not uncommon for shifts in ostensible topic to occur in this manner. As I have noted elsewhere (Horlick-Jones, 2005b), such groups appear to be engaged in some kind of negotiation process, with speakers taking the opportunity to offer interpretations that reflect their own perspectives, experiences and moral commitments.

At line 27, M4 reports on mediated knowledge he has accessed which suggests that mobile phones 'heat the brain cells'. Despite this possibly alarming information, he adds 'you've just got to believe what they tell you about it they say they're not dangerous'. What is going on here? Whether his latter comment indicates a general need to believe authoritative sources ('they'), or a need to not worry in order to 'get though the day', is not clear. Importantly, M4 can be heard as distancing himself from responsibility for managing possible risks associated with mobile telephone use. M1 and F2 then rein-forces this sentiment by together using another lay logical device at lines 30–35 (punctuated by the group moderator seeking clarification

Figure 3.3   A second illustration of the dynamics of talk about GM issues

NBD/4/12-13 (M$_i$ = male participants; F$_i$ = female participants; Mod = moderator)		
27	M4	just carry on using it I saw a documentary once they … heat the brain cells up and
28		whatever and you just got to believe what they tell you about it they say they're
29		not dangerous
30	M1	there's so many things you can't do and should do you shouldn't do this if you
31		listened to everything who wouldn't do
32	F2	you wouldn't go out the front door again
33	Mod	sorry ((F2))
34	F2	walk through your front door and get run over that would be it so do you listen to
35		things or just
36	F3	it's moderation isn't it
37	M3	use your judgement I mean I keep my mobile phones away from my son I never
38		put it near him but um because of things I've heard but as far as using it myself I
39		use it all the time

Adapted from Horlick-Jones *et al.*, 2007b.

at line 33) which says essentially that 'you wouldn't do anything if you listened to all these warnings'.

The use of the 'you wouldn't do anything' device, followed at line 36 by F3's re-iteration of the 'moderation' argument, have an essentially fatalist message, which appear to be used in an to attempt to lessen possible anxieties associated with mobile phone use, and to diminish associated culpabilities. Indeed, in the following lines 37–39, M3 observes that despite his own heavy use of his mobile phone, he keeps it away from his son. In this way he may be heard as portraying himself in a 'good light' as a parent in the context of the earlier discussion. In this way, social accountability may be seen to play an important role in the dynamics of the group's attempts to make sense of the technology of genetic modification and its associated risks. This observation chimes with the claim that risk-related issues have the capacity to bring out social accounting practices in particularly forceful ways, requiring actors to account for risk-related actions in ways that not only make sense, but also presents them in morally acceptable ways (Horlick-Jones, 2005a; 2005b).

Although much of the groups' discussions featured possible concerns about GM, there were a number of instances when the participants became quite excited by possibilities that the technology seemed to make possible. This tendency is illustrated by the sequence of talk reproduced at Figure 3.4. At lines 1–3, M2 introduces the idea of ice cream produced from genetically modified components which would be very low in calories, and therefore not fattening. The group is much amused and animated by this intervention. F3 quickly introduces the

Figure 3.4   An example of playful and imagination talk about GM issues

NBD/4/53(1) (M$_i$ = male participants; F$_i$ = female participants; Mod = moderator)		
1	M2	I mean you could have all sorts of food, I suppose you don't put on any weight
2		on, you know, you could eat as much ice cream as you like cause it's got one
3		calorie per tub
4	F3	marvellous!
5		((Laughter))
6		((data gap))
7	Mod	so who's up for it? ... hang on ... it's a great question
8	F3	GM chocolate, definitely!
9	Mod	GM chocolate, yeah
10	M2	but chocolate that had no calories ... you could have, like, six bars and be, like,
11		that ... it tastes the same ... if Cadbury's said 'oh we've got this little GM lab going
12		on to see', you know
13	Mod	would you buy it?
14	M2	yeah, if it didn't have an adverse effect
15	F2	it's an ideal thing, isn't it?
16	Mod	quick show of hands ... we've got F4 and F2
17	M2	See, we knew we'd win over with chocolate!
18		((Laughter))

Adapted from Horlick-Jones *et al.*, 2007c.

idea of 'GM chocolate', and M2 develops this idea by suggesting that this chocolate might also have non-fattening properties. In response to the apparent excitement of participants, the moderator seeks to probe the degree of unanimity among the group. It become clear that the whole group is similarly enthusiastic about what F2 describes as 'an ideal thing'.

A number of these 'GM fantasy products' were 'invented' during the various group discussions, including 'GM alcohol that doesn't give you a hangover' and 'GM tobacco that doesn't kill you' (Horlick-Jones *et al.*, 2007b). This activity illustrates the degree of playful inventiveness that can characterise such processes of sense-making; as the groups interrogated the incomplete knowledge they possessed, and creatively explored possibilities that related to matters of importance to them.

All new technologies bring with them the possibility of unintended adverse consequences. The recent experience in Britain of BSE and other food scares strongly reinforces this caution. Lay people might then quite reasonably ask of genetic modification 'what's in it for me?', and the answer is not at all clear. In contrast, people clearly see themselves as gaining significant benefits from using mobile telephones, and the warnings about adverse health consequences appear to have had little impact on the widespread use of this technology (Walls *et al.*, 2005). The implication would appear to be that caution about genetic modification might similarly be swept aside if suitably attractive GM-related commodities became available. Perhaps time will tell.

Concluding remarks

In developing the approach set out in this chapter[2] to analysing, and empirically investigating, the public acceptability of technologies in terms of the concept of signature, I referred back to some earlier writing about technology in which I argued that (Horlick-Jones, 1996: 145): 'Technical, human, managerial and cultural dimensions interact in a contingent open-ended process that precludes deterministic analysis'. I have attempted to capture a similar sense of open-endedness and dynamism here. The market take-up of, or resistance to, new technologies emerges from just such a multidimensional field of phenomenal relations to which actors orientate. In that sense, for a given social group, a technology's signature will reflect its material nature, and their circumstances, preferred ways of life and cultural sensibilities. It will also reflect their degree of awareness and understanding of the issues in question, which, by necessity, will be incomplete. Far from being a fixed entity, a signature may change with the shifting interaction between a multitude of constraining and enabling factors, including economics, media portrayals, marketing, and oppositional campaigns, as they evolve in time.

My development of the notion of signature has drawn upon a broadly ethnomethodological perspective: one that has increasingly informed my thinking about risk, action and experience in recent years. It has also been influenced by a concern with the roles of knowledge and social accounting in shaping distinctive reasoning practices associated with risk-related issues. As such, I have recognised the importance of the local, informal logics that emerge from social interaction as actors fashion the social world as a practical ongoing accomplishment in mutually-intelligible and morally accountable ways. Such interaction is in turn shaped and constrained by a host of shifting social and material influences. Social resistance to (or acceptance of) technologies is, in this sense, 'constructed' (or, as ethnomethodologists would prefer to put it, 'produced': Lynch, 2001) within the fabric of that reflexive dynamic.

Conventional constructionist accounts provide an important source of enrichment for debates about the social acceptability of technology, by moving them away from a narrow technocratic reductionism. However in this chapter I have argued that such accounts have achieved these gains at the risk of losing track of the specific features of the very techno-logical artefacts that they address. The danger here lies in the incorpora-tion of sociological theories of reality (and rationality) into the analysis, at the expense of engaging with the specific ways in which technologies are articulated in practical reasoning and mundane discourse within real-world settings. It would seem that such features of how technologies are apprehended would play a crucial role in shaping their social acceptabil-ity. Ultimately, of course, these are matters for empirical investigation.

Acknowledgements

The ideas discussed in this chapter draw upon many invaluable conversa-tions with colleagues, and I am delighted to take this opportunity thank some of them: Julie Barnett, Paul Bellaby, Mick Bloor, Adam Burgess, Rob Flynn, Ian Hutchby, Jenny Kitzinger, Graham Murdock, Greg Myers, Christian Oltra, Judith Petts, Ana Prades López, Gene Rowe and John Walls. I am also pleased to acknowledge the role of the late Jonathan Sime, with whom I developed a number of the arguments used here.

Notes

1. In making this observation, it may appear that I am being a little unfair to Burgess, whose work on mobile telecommunications, I am pleased to acknowledge, is notable for its detailed examination of economic, political, cultural and historical contextual factors. My concern here is that Burgess seems to assume that lay people will act in simple compliance with the 'inexorable logic' (Burgess, 2004: 91) of the 'culture of precaution' that is generated by these contextual factors. The arguments I present in this chapter suggests that in the real world, lay reasoning about technologies and their associated risks is rather more complicated in nature. I note that recent empirical investigations of the public understanding of precautionary regula-tion in the context of mobile telephones (Timotijevic and Barnett, 2006, and Chapter 5 in this volume) indicates the need for just such a nuanced under-standing of lay reasoning.
2. I am only too aware that in setting out these arguments I have glossed over a number of methodological matters. In particular, whilst my analytical approach demands an attention to data, which is, in some sense, naturalis-tic, and to the study of 'naturally organised ordinary activities' (e.g. Lynch, 1993), the reader will note that here I have depended heavily upon data gen-erated by focus groups. I have addressed this matter in detail elsewhere (Horlick-Jones *et al.*, 2007b).

References

D. Anderson and P. Mullen (eds), *Faking It: the Sentimentalisation of Modern Society* (Harmondsworth: Penguin, 1998).

R. Anderson and W. Sharrock, 'Can organisations afford knowledge?', *Computer Supported Cooperative Work*, 1 (1993) 143–161.

B. Barnes, *The Elements of Social Theory* (London: UCL Press, 1995).

U. Beck, *Risk Society: Towards a New Modernity* (London, Sage, 1992).

P. Bellaby, 'Communication and miscommunication of risk: understanding UK patient's attitudes to combined MMR vaccination', *British Medical Journal*, 327 (2003) 725–728.

P. Berger and T. Luckman, *The Social Construction of Reality* (Harmondsworth: Penguin, 1996).

J. Best (ed.), *Images of Issues: Typifying Contemporary Social Problems* (New York: Aldine De Gruyter, 1989).

W. Bijker and J. Law (eds), *Shaping Technology/Building Society: Studies in Sociotechnical Change* (Cambridge, Mass: MIT Press, 1992).

M. Bloor, R. Datta, Y. Gilinskiy and T. Horlick-Jones, 'Unicorn among the cedars: on the possibility of effective "smart regulation" of the globalized shipping industry', *Social & Legal Studies* (2006) 15, 4, 534–551.

D. Boden, 'Worlds in action: information, instantaneity and global futures trading', in Adam, B., Beck, U. and Van Loon, J. (eds) *The Risk Society and Beyond: Critical Issues for Social Theory* (London: Sage, 2000) 183–197.

A. Burgess, *Cellular Phones, Public Fears and a Culture of Precaution* (Cambridge: Cambridge University Press, 2004).

K. Burningham and G. Cooper, 'Being constructive: social constructionism and the environment', *Sociology*, 33, 2 (1999) 297–316.

G. Button (ed.), *Ethnomethodology and the Human Sciences* (Cambridge: Cambridge University Press, 1991).

G. Button, 'Introduction' and 'The curious case of the vanishing technology', in Button, G. (ed.) *Technology in Working Order: Studies of Work, Interaction and Technology* (London: Routledge, 1993), 7–28.

C. Candlin and S. Candlin, 'Discourse, expertise and the management of risk in healthcare settings', *Research on Language and Social Interaction*, 35 (2002) 115–137.

CEC (2001) *European Governance: a White Paper*, Commission of the European Communities, Brussels.

D. Collingridge, *The Management of Scale: Big Organizations, Big Decisions, Big Mistakes* (London: Routledge, 1992).

R. Dingwall, '"Risk Society": the cult of theory and the millennium?', *Social Policy & Administration*, 33, 4 (1999) 474–491.

M. Douglas, 'Risk as a forensic resource', *Daedalus*, 119, 4 (1990) 1–16.

M. Douglas and A. Wildavsky, *Risk and Culture: an Essay on the Selection of Technological and Environmental Dangers* (Berkeley: University of California Press, 1982).

B. Durodié, 'The true cost of precautionary chemical regulation', *Risk Analysis*, 23, 2 (2003) 389–398.

N. Fox, '"Risks", "hazards" and life choices: reflections on health at work', *Sociology*, 32, 4 (1998) 665–687.

F. Furedi, *Culture of Fear: Risk-Taking and the Morality of Low Expectations* (London: Cassell, 1997).

H. Garfinkel, *Studies in Ethnomethodology* (Englewood Cliffs NJ: Prentice-Hall, 1967).

J.J. Gibson, *The Ecological Approach to Visual Perception* (Boston: Houghton Mifflin, 1979).

A. Giddens, *Modernity and Self-Identity* (Cambridge: Polity, 1991).

B. Glassner, *The Culture of Fear: why Americans are Afraid of the Wrong Things* (New York: Basic Books, 1999).

I. Hacking, *The Social Construction of What?* (Cambridge, Mass: Harvard University Press, 1999).

J. Hannigan, *Environmental Sociology: A Social Constructionist Perspective* (London: Routledge, 1995).

S. Hilgartner, 'The social construction of risk objects: or, how to pry open networks of risk', in Short, J.F. Jr. and Clarke, L. (eds) *Organizations, Uncertainty, and Risk* (Boulder, Col.: Westview Press, 1992) 39–53.

C. Hood and D. Jones, 'Preface', in Hood, C. and Jones, D. (eds) *Accident and Design: Contemporary Debates in Risk Management* (London: UCL Press, 1996) xi–xiii.

T. Horlick-Jones, 'Is safety a by-product of quality management?', in Hood, C. and Jones, D. (eds) *Accident and Design: Contemporary Debates in Risk Management* (London: UCL Press, 1996) 144–154.

T. Horlick-Jones, 'Science – the language of the powerful?', *Journal of Risk Research*, 1, 4 (1998a) 321–325.

T. Horlick-Jones, 'Meaning and contextualisation in risk assessment', *Reliability Engineering & System Safety*, 59 (1998b) 79–89.

T. Horlick-Jones, 'Experts in risk?...do they exist?', *Health, Risk & Society*, 6, 2 (2004) 107–114.

T. Horlick-Jones, 'On "risk work": professional discourse, accountability and everyday action', *Health, Risk & Society*, 7, 3 (2005a) 293–307.

T. Horlick-Jones, 'Informal logics of risk: contingency and modes of practical reasoning', *Journal of Risk Research*, 8, 3 (2005b) 253–272.

T. Horlick-Jones and B. De Marchi, 'The crisis of scientific expertise in *fin de siècle* Europe', *Science and Public Policy*, 22, June (1995) 139–145.

T. Horlick-Jones and J. Sime, 'Living on the border: knowledge, risk and trans-disciplinarity', *Futures*, 36 (2004) 441–456.

T. Horlick-Jones, J. Walls, G. Rowe, N. Pidgeon, W. Poortinga, G. Murdock and T. O'Riordan, *The GM Debate: Risk, Politics and Public Engagement* (London: Routledge, 2007a).

T. Horlick-Jones, J. Walls and J. Kitzinger, '*Bricolage* in action: learning about, making sense of, and discussing issues about GM crops and food', *Health, Risk & Society* (2007b).

T. Horlick-Jones, G. Rowe and J. Walls, 'Citizen engagement processes as information systems: the role of knowledge and the concept of translation quality', *Public Understanding of Science* (London: HMSO, 2007c) 16, 3, 259–278.

House of Lords Select Committee on Science and Technology, Third Report, 'Science and Society', HL Paper 38 (London: HMSO, 2000).

P. Huber, *Galileo's Revenge: Junk Science in the Courtroom* (New York: Basic Books, 1991).

I. Hutchby, *Conversation and Technology: from the Telephone to the Internet* (Cambridge: Polity, 2001).

A. Irwin and B. Wynne (eds) *Misunderstanding Science: the Public Reconstruction of Science and Technology* (Cambridge: Cambridge University Press, 1996).

M. Kaldor, *The Baroque Arsenal* (London: Andre Deutsch, 1982).

S. Lash, B. Szerzynski and B. Wynne (eds) *Risk, Environment & Modernity: Towards a New Ecology* (London: Sage, 1996).

R. Löfstedt and T. Horlick-Jones, 'Environmental politics in the UK: institutional change and public trust', in Cvetkovich, G. and Löfstedt, R. (eds) *Social Trust and the Management of Risk* (London: Earthscan, 1999) 73–88.

R. Löfstedt and O. Renn, 'The Brent Spar controversy: an example of risk communication gone wrong', *Risk Analysis*, 17, 2 (1997) 131–136.

D. Lupton, *Risk* (London: Routledge, 1999).

M. Lynch, *Scientific Practice and Ordinary Action: Ethnomethodology and Social Studies of Science* (Cambridge: Cambridge University Press, 1993).

M. Lynch, 'The contingencies of social construction', *Economy and Society*, 30, 2 (2001) 240–254.

D. Maynard, *Bad News, Good News: Conversational Order in Everyday Talk and Clinical Settings* (Chicago: University of Chicago Press, 2003).

H. Molotch, *Where Stuff Comes From: How Toasters, Toilets, Cars, Computers, and Many Other Things Come to Be as They Are* (New York: Routledge, 2003).

M. Mulkay, Review of *Misunderstanding Science? The Public Construction of Science and Technology*, in Irwin, A. and Wynne, B. (eds) *Science, Technology & Human Values*, 22, 2 (1997) 254–264.

G. Myers and P. Macnaghten, 'Rhetorics of environmental sustainability: commonplaces and places', *Environment and Planning A*, 30 (1998) 333–353.

C. Norris, 'Truth, science and the growth of knowledge', *New Left Review*, 210 (1995) 105–123.

OECD, *Engaging Citizens in Policy-Making: Information, Consultation and Public Participation*, PUMA Policy briefing No. 10 (Paris: Organisation of Economic Co-Operation and Development, 2001).

T. O'Riordan, R. Kemp and M. Purdue, *Sizewell B: an Anatomy of the Inquiry* (Basingstoke: Macmillan, 1988).

C. Perrow, *Normal Accidents: Living with High-Risk Technologies* (New York: Basic Books, 1984).

J. Petts, T. Horlick-Jones and G. Murdock, *Social Amplification of Risk: the Media and the Public* (Sudbury: HSE Books, 2001).

Lord Phillips of Worth Matravers, J. Bridgeman and M. Ferguson-Smith, *The BSE Inquiry ('The Phillips Inquiry')* (London: HMSO, 2000).

A. Pickering (ed.), *Science as Practice and Culture* (Chicago: University of Chicago Press, 1992).

O. Renn, 'Concepts of risk: a classification', in: Krimsky, S. and Golding, D. (eds) *Social Theories of Risk* (Westport, Conn: Praeger, 1992) 53–79.

N. Rose, 'Authority and the genealogy of subjectivity', in Heelas, P., Lash, S. and Morris, P. (eds) *Detraditionalization: Critical Reflections of Authority and Identity* (Oxford: Blackwell, 1996) 294–327.

N. Rose, *Powers of Freedom: Reframing Political Thought* (Cambridge: Cambridge University Press, 1999).

H. Rothstein, 'The institutional origins of risk: a new agenda for risk research', *Health, Risk & Society*, 8, 3 (2006) 215–221.

Royal Society, 'The Public Understanding of Science' (London: Royal Society, 1985).

A. Sayer, *Realism and Social Science* (London: Sage, 2000).

A. Sayer (2004) 'Restoring the moral dimension', Department of Sociology, University of Lancaster: http;//www.comp.lancs.ac.uk/sociology/papers/sayer-restoring-moral-dimension.pdf

J. Shotter and K. Gergen (eds), *Texts of Identity* (London: Sage, 1989).

J. Sime, 'What is environmental psychology? Texts, content and context', *Journal of Environmental Psychology*, 19 (1999) 191–206.

P. Slovic, *The Perception of Risk* (London: Earthscan, 2000).

M. Spector and J. Kitsuse, *Constructing Social Problems* (New York: Aldine De Gruyter, 1987).

P. Strong, 'Sociological imperialism and the profession of medicine: a critical examination of the thesis of medical imperialism', *Social Science & Medicine* 13A (1979) 199–215.

P. Strong and R. Dingwall, 'Romantics and Stoics', in Gubrium, J. and Silverman, D. (eds) *The Politics of Field Research: Sociology Beyond Enlightenment* (London: Sage, 1989) 49–69.

L. Timotijevic and J. Barnett, 'Managing the possible health risks of mobile telecommunications: public understanding of precautionary action and advice', *Health, Risk & Society*, 8, 2 (2006) 143–164.

I. Velody and R. Williams (eds), *The Politics of Constructionism* (London: Sage, 1998).

J. Walls, N. Pidgeon, A. Weyman and T. Horlick-Jones, 'Critical trust: understanding lay perceptions of health and safety risk regulation', *Health, Risk & Society*, 6, 2 (2004) 133–150.

J. Walls, T. Horlick-Jones, J. Niewöhner, and T. O'Riordan, 'The meta-governance of risk: GM crops and mobile telephones', *Journal of Risk Research*, 8, 7–8 (2005) 635–661.

S. Woolgar and D. Pawluch, 'Ontological Gerrymandering: the anatomy of social problems explanations', *Social Problems*, 32 (1985) 214–227.

B. Wynne, *Rationality and Ritual: the Windscale Inquiry and Nuclear Decisions in Britain* (Chalfont St. Giles: The British Society for the History of Science, 1982).

B. Wynne, 'SSK's identity parade: signing-up, off-and-on', *Social Studies of Science*, 26 (1996) 357–391.

B. Wynne, 'Creating public alienation: expert cultures of risk and ethics on GMOs', *Science as Culture*, 10, 4 (2001) 445–481.

4
Public Acceptance of New Technologies in Food Products and Production

Arnout R.H. Fischer and Lynn J. Frewer

Introduction

In this chapter, we will discuss the public acceptance of new technologies from the perspective of food production and food products. Food is of particular interest in this context as it serves to illustrate many of the relevant issues pertinent to the introduction and application of emerging technologies more generally. Many food products are produced using traditional methods and approaches. However, food products can also be developed using innovative technologies which, furthermore, may be linked to new qualities or attributes in food products. In addition, food consumption is not only a biological necessity, it is also part of people's lives, and is associated with cultural and social significance, as well as pleasurable or unpleasant sensory experiences. Thus people's responses to food are not only based on their assessment of its nutritional characteristics, but on various attributes including quality, social context, and hedonistic response. For example, among the wide range of food products that are produced using traditional and long established methods, many are valued by consumers on the basis of perceived naturalness and application of organic or artisanal production methods (Van Rijswijk *et al.*, in preparation). This is reflected by the introduction of, for example, authenticity labelling by institutions such as the European Commission.[1] The importance of social context implies that efforts to introduce novel foods or the food products of new technologies, without a broader understanding of the factors underpinning people's food choices, may result in consumer rejection of both food products themselves, and the technology which is used to produce them. This has been illustrated in the case of genetically modified (Frewer *et al.*, 2004) or irradiated foods (Bruhn, 1995).

To understand consumer behaviour towards novel food, it is important to have an idea how consumers respond to novel foods and ingredients in comparison to their reactions to traditional foods.

Food production through traditional and novel methods

One reason to adopt novel technologies is that these may, to some extent, be used to mitigate the risks associated with food hazards, as well as deliver benefits in terms of improved nutritional quality or more sustainable production. Although many improvements in food safety of traditional products have been achieved in recent decades, microbial, toxicological, and carcinogenic substances are still sometimes found in products destined for human consumption (Kreijl *et al.*, 2004). In addition, human health may be compromised by inappropriate nutrition linked to dietary choices or over-consumption of specific food components. As a consequence, there has been increased emphasis on promoting healthy eating to reduce cardiovascular problems and obesity in developed countries. Novel applications of biotechnology which have been developed in order to produce plants with advantageous traits have the potential to improve quality of life in society. Various applications have been developed which confer benefits in terms of human nutrition and micronutrient delivery, bio-security and development of varieties which grow in hostile environmental conditions, are resistant to pests or pesticides, or which have other desirable qualities such as improved aesthetic presentation.

However, the introduction of new products and technologies may also introduce new potential hazards to the food chain, such as allergic reactions to novel proteins (Van Putten *et al.*, 2006, Kuiper *et al.*, 2002) or negative environmental impacts. The potential risks and benefits associated with such new developments may not be fully recognised by the general public. Even among experts, uncertainty associated with risk-benefit judgements may exist in the context of consumer protection. Under these circumstances, trust in risk regulators, food producers and scientists is important if emerging food technologies are to be accepted by society (Eiser *et al.*, 2002, Siegrist, 2000). Consequently, food can be described as a 'hazard domain' where both people's lifestyles and attitudes to new technologies need to be taken into account when considering how consumers will react to different foods and food ingredients. Other such 'domains' may be identified which also include societal concerns about existing and emerging production technologies.

For example, in the energy sector, there has been a range of energy producing techniques introduced subsequent to the widespread societal adoption of electricity produced using fossil fuels. Nuclear energy production (Slovic *et al.*, 1991) and, more recently, wind, solar and biomass have been introduced as new production technologies. As these differ in perceived novelty and level technology they differentially lead to a greater or lesser degree of societal concern (van den Hoogen *et al.*, 2006). More recently, technological developments have raised the possibility of introducing hydrogen as energy carrier (Solomon and Banerjee, 2006). Some of these domains and applications of new technology have shown, or are showing, sufficient similarities to the food sector to identify common elements which may be important when considering potential consumer acceptance (Frewer *et al.*, 2003). One of the common elements is that the discussions associated with the development and commercialisation of new technologies have been frequently conducted in communities (mainly) consisting of experts drawn from the natural and technological sciences. In the past, expert groups have criticised negative consumer attitudes towards some food technologies, (for example, genetic engineering applied to food production), while failing to consider the origins of these consumer attitudes. Such 'expert' responses to consumer concern have frequently been contextualised by the observation that consumers accept exposure to potentially larger risks through unhealthy food choices and other potentially hazardous food consumption patterns. The behaviour of consumers in relation to food safety issues can, however, only be properly understood if there is systematic understanding of the way in which consumers perceive risks, and indeed benefits, and how these relate to an effective food safety and technology development and commercialisation strategy (Frewer and Salter, 2003).

Food choice and the psychology of risk

Within the area of risk psychology, much effort has been invested in identifying and quantifying underlying factors in consumer responses to situations, which has been termed the psychometric approach to understanding risk perception. With regard to risks, it has been demonstrated that relevant psychological factors determine people's responses to a particular hazard (Fischhoff *et al.*, 1978; Slovic, 1987; Slovic, 1993). Among the most important findings was the observation that factors that are not included in technical risk estimates may influence people's perception of risk, such as the extent to which a risk is perceived to be unnatural, potentially catastrophic, or in which exposure is involuntary. These psychological dimensions are excellent predictors of people's responses across *different* hazard domains, including food hazards (Fife-Schaw and Rowe, 2000). People's concerns may be very specific to particular hazard domains, including that of food and food production technologies (Frewer, 2003).

In the area of food and technology acceptance, the perception that a particular technology may potentially have a negative impact on nature, or other areas or values that people consider within the natural world, is an important determinant of consumer responses to the technology under consideration and its products (Miles and Frewer, 2001). For example, in the case of Bovine Spongiform Encephalopathy (BSE) people may be concerned about the potentially detrimental effect of the disease on animal welfare, which may not apply to other types of potential hazard (Miles and Frewer, 2001). Some consumers appear to be especially *neophobic* in terms of their reactions to novel foods (e.g. Tuorila *et al.*, 2001; Bredahl, 2001), a human response which may have evolved in order to protect people from consuming potentially toxic new foods (Rozin and Vollmecke, 1986). There is evidence that neophobia is generally greater for older people, and reduces as educational level increases (Tuorila *et al.*, 2001). Research is, however, needed to determine whether neophobia inherently increases with age, or whether it will gradually disappear as the younger people, who are at present more positive toward new technology, become the older consumers of tomorrow.

Public attitudes and risk perception

Public perceptions and attitudes about emerging bio-sciences and other new technologies applied to food production are among the most important factors determining the likelihood of the successful development and implementation of agri-food technologies (Frewer *et al.*, 2004). Differences in risk perceptions associated with lifestyle hazards on one hand, and technological food-related hazards on the other, have been identified (Fife-Schaw and Rowe, 2000; Miles *et al.*, 2004).

Other factors include 'optimistic bias' (Weinstein, 1989), where consumers perceive that they are at less risk than a member of society with whom they compare their own risks. Optimistic bias is more commonly observed for lifestyle food related hazards, where people perceive they have higher levels of personal control over hazard exposure compared to more vulnerable, and less knowledgeable others. Optimistic bias tends to disappear when the potential hazards are technological in origin, and hence perceived to be less amenable to personal control (Frewer *et al.*, 1994). However, repeated exposure of an individual to a potentially hazardous situation leads to a strong and stable risk attitude, which is not easily changed by risk communication (Fischer and De Vries, submitted). It is possible that optimistic bias arises under circumstances when repeated exposure does not immediately lead to negative consequences, and people perceive they have a high level of personal control over their exposure. Once an optimistic bias has formed, it appears to be relatively stable. For example, there is evidence that consumers who have actually suffered from a food-induced illness show only a temporal diminishing of their optimistic bias towards microbial food safety (Parry *et al.*, 2004). In many lifestyle situations optimistic biases mean that consumers *underestimate* their personal risks from a particular hazard.

In contrast to lifestyle hazards, consumers may react negatively to the introduction of specific food technologies such as food irradiation and gene technology, in such a way that consumers *overestimate* the risks associated hazards levels compared to the estimates provided by experts. It has been argued that the ways technical risk experts and lay people think about the risks associated with different technical applications are very different (e.g. Lazo *et al.*, 2000; Mertz *et al.*, 1998; Slovic *et al.*, 1995). In general, consumers appear more concerned about risks which are related to the development and application of technology in comparison to naturally occurring risks, even when there is an equal probability of harm to human health (Hansen *et al.*, 2003). Consumers are often seen as non-rational decision-makers by experts. If it is, however, taken into account that a logical weighing of arguments requires substantial mental resources, attention, and motivation it can be understood why consumers do not always follow the so-called rational arguments of experts.

Dual-process models of attitude change

Theories of persuasion have been developed in order to understand why, and under what circumstances, information may change people's attitudes regarding a particular issue, and to understand why differences in persuasion may occur across different consumers and information domains. That is, not all participants in all situations react the same way to persuasive argumentation. *Dual-process* models of attitude change have attempted to explain the situational and contextual circumstances under which people change their attitudes following presentation of relevant information. It is now generally accepted that cognitive effort is required to process information in the in-depth and thoughtful way (systematic or elaborate processing) needed to induce attitude change (Cacioppo *et al.*, 1986). An individual needs to be motivated in order to expend this effort, and an individual must also possess, and be willing to expend, the mental resources to process the information. Under circumstances where only limited cognitive effort is available to process the information, heuristics (or decision-rules) and other short cuts in reasoning are applied to reduce the effort which needs to be expended in the information processing task. This type of information processing is termed 'heuristic processing'.

When heuristic processing is applied, attitude changes are less predictable and stable compared to situations where elaborate or systematic processing has been applied. Modern versions of the elaboration likelihood model (Petty and Wegener, 1999) and the heuristic-systematic model (Chen and Chaiken, 1999), assume that elaborate and heuristic processing will often both occur during the processing of provided information. If we assume that consumer acceptance or rejection of new food technologies should be based on the best and most balanced information available, elaborate or systematic processing on the part of consumers should be the dominant path of information processing. If information can be made highly relevant to the person receiving the information, their motivation to process this information in an elaborate way will be increased (Fazio and Towles-Schwein, 1999). A successful approach may be to design the information in such a way that a heuristic cue communicates the personal relevance to the recipient of the information, resulting in subsequent elaborate processing. Thus in terms of risk communication, it is of relevance to know what conditions allow the systematic processing of information; and what heuristics activate elaborate processing.

Persuasive communication theories provide an infrastructure in which attitude change might be induced. In other words, the focus of this approach is to convince the consumer to adopt a specified point of view. However, consumer trust might best be developed by the provision of transparent information about the risks and benefits and regulation of new products (Houghton *et al.*, in press). Transparent communication implies that different points of view are presented, and that both risk and benefits are communicated, although this may introduce an ambivalent message. In the past, the solution to such unpredictability was to avoid presentation of mixed messages.

However, if attitudes have been established, and the message does not support the existing attitudes held by an individual, increased distrust in the message source may occur, whilst at the same time attitudes remain unchanged (Frewer *et al.*, 2003).

Furthermore, it has been argued that the weighing of the positive and negative aspects of a particular hazard, and making trade-offs between them, is an inherently subjective action (Slovic, 1999). In addition, consumers have increasing access to different sources of information about a potentially controversial topic (for example, via the Internet) and so are no longer reliant on the views of experts (Frewer and Salter, 2002). Providing potentially ambiguous messages to the public assumes that experts accept that the consumer may differentially assign weights to risks and benefits to arrive at a conclusion regarding the acceptability or otherwise of a specific hazard, which may result in individual members of society expressing opinions that do not necessarily align with expert judgements. Providing consumers with the informational 'tools' needed to make informed choices about emerging technologies is needed if effective consumer involvement in technology development and regulation is to occur (Rowe and Frewer, 2000).

Heuristic processing of information

One of the reasons why consumer perceptions differ from those of experts, may be the different use of heuristics in processing the available information. The role of *heuristics* in human information processing has represented an important focus of research in recent decades. The realisation has grown that many human information processing mechanisms may be underpinned by heuristics, which can be applied by an individual receiving new information to reduce the amount of effort needed to process new information. Such an approach can be very effective and efficient, at least when such an approach results in the same conclusions as a systematic weighing of arguments. This may facilitate reduction of the complexity of the information environment in which an individual finds himself or herself. One way to reduce the amount of effort is by adopting the opinion of another person or group of people regarding a particular subject or topic. This may indeed be a very efficient strategy if an individual believes that the other person has systematically appraised the different attributes of a given topic, and, at the same time, possesses the expertise needed to judge the merits or otherwise of the information.

Thus the extent to which a source is perceived to possess *expertise* may act as a cue that increases the likelihood of persuasion occurring, regardless of underlying arguments (McGuire, 1985). Perceived honesty and lack of vested interest associated with promoting a particular view may also contribute to persuasion (e.g. Frewer *et al.*, 1996), although honesty without expertise may not have value in this respect. Nonetheless, if information is provided by a trusted source, acceptance of the conclusions may occur, independent of the argumentation provided in support of those conclusions. However, these effects may also be dependent of the type of hazard under consideration. For example, in the case of communication about microbial food safety, there is evidence that information source characteristics are less influential than *message relevance* in influencing risk perceptions associated with food poisoning (Frewer *et al.*, 1997a).

Prior attitudes, personal experience and even automated behavioural patterns such as habits may also serve as heuristic determinants of behaviour. Particularly in the case of routine tasks like food preparation, one might expect that mental processes such as scripts or habits are very important (Fischer and De Vries, submitted). Even for relatively new technologies such as genetically modified organisms, the influence of existing attitudes on new (persuasive) information about the technology is shown to be important (Eiser *et al.*, 2002; Frewer *et al.*, 1998), as more extreme attitudes are less likely to be amenable to change through presentation of contradictory but persuasive information.

Emotions, risk and attitude change

Another heuristic process, the *affect heuristic* (Slovic *et al.*, 2004), has been derived from the observation that risk and benefit perceptions are, in general, negatively correlated (Alhakami and Slovic, 1994). This emotion-related heuristic implies that when an individual is experiencing positive emotions about a specific activity or event, the risks associated with the activity or event will be perceived as *lower* and the associated benefits as *higher* (Finucane *et al.*, 2000). More generally, one might predict that positive mood results in a shift towards a more positive attitude if the information provided matches the direction of attitude change, whereas negative affect has the converse effect on persuasion (Slovic *et al.*, 2004).

This also suggests that applying certain emotions in risk communication may facilitate successful processing of information. Fear about potential ill-effects associated with an event or behaviour has often been used to motivate consumer to process arguments systematically in order to achieve attitude changes (Witte and Allen, 2000), although the results of research where fear has been applied as a potential motivator have been somewhat equivocal (Ruiter *et al.*, 2001). One of the problems with using emotional cues embedded in, or associated with, a persuasive message is the possibility that emotions in themselves represent heuristic cues. If the motivational effect of a specific emotion is not aligned with the message itself, the effect of applying emotions may lead to unexpected effects (see e.g. Meijnders *et al.*, 2001). For example, fear may primarily activate self-protective behaviour, regardless of the message that is being communicated.

The exact roles of emotions such as heuristics in processing persuasive information are not well understood at the present time. Many researchers in communication science and social psychology focus on the distinction between positive and negative emotions. However, some researchers argue that emotions can have specialised functions (Lerner and Keltner, 2000; Lerner and Keltner, 2001). For example, it is postulated that anger, or aggression, is an emotion that mobilises resources to fight out of a problematic situation, whilst at the same time temporarily disregarding personal safety and short term goals relating to self-protection (Oatley and Johnson-Laird, 1987). It is arguable that, in certain situations, short term disregard for self-protection may result in long term benefits, e.g. by standing up against aggressors to show that you are not tolerant of the aggression expressed by them. The long term positive effect, a lower probability of experiencing threat from aggressors, will compensate for the short term negative effect of experience of physical injury (Evans, 2002).

On the other hand, disgust is an emotion that provides a signal designed to prevent close contact with the object which produced the disgust in the first place. For example, in the context of food consumption, disgust may prevent people consuming potentially harmful foods (Rozin and Fallon, 1987). In a recent study, it has been shown that when communication about preventing food-borne diseases was accompanied by relevant images designed to invoke disgust, participants subsequently used the information provided as a means to achieve safer food handling behaviours. When the same messages were accompanied by aggressive images, the efficacy of the message was very much reduced, to the extent that it was even less effective than a version of the same message in which no emotional images were provided (Nauta *et al.*, in preparation). This provides evidence that different negative emotions need not have the same effect on how people use information. Achieving a better understanding of the effects of emotions on attitudes and behaviour is an increasingly important research area in social psychological research.

In summary, the different angles from risk psychology on communication about foods imply that:

1) Two *psychometric dimensions*, newness (new technology, risks unknown to science) and dreadedness (impact on nature, number of people affected, voluntary exposure) predict consumer risk perceptions. Specific concerns may emerge directly linked to particular hazard domains, and consumers may be neophobic with regard to novel foods. Thus, for new technological products consumers tend to assess risks as being higher than would be predicted by the technical risk assessments provided by experts.

2) *Ethics and values* may play an important role in consumer acceptance of technology and its applications in certain situations.

3) For *lifestyle hazards*, consumers tend to arrive at lower risk perception than risk estimates by experts. This lower relative level of consumer risk perception may be due to personal experience of the hazard, or optimistic bias regarding one's own risk from the hazard. An additional barrier to changing behaviour within this type of hazard domain may be the tendency to develop habitual responses to frequently performed behaviours.

4) Central or *systematic processing* of persuasive information is essential if lasting change in attitudes is to occur. To achieve systematic processing, the individual receiving the information needs to be motivated to process the information, as well as possessing sufficient cognitive resources to process the new material.

5) *Heuristic processing* may contribute to attitude change. *Trust, prior experience* and *emotion* may all play important roles in determining how people respond to persuasive information, but these different areas are worthy of further research.

At this point, it is useful to consider the above in the context of a specific case study relevant to the introduction of emerging technologies, the introduction of Genetically Modified (GM) foods into the European Union.

The introduction of GM foods in Europe

In the mid-1990s genetically modified food products were introduced into the European food markets. In the latter part of 1996, controversy over gene technology in Europe had become widespread, triggered by the arrival of non-segregated genetically modified soybeans in European harbours (Lassen *et al.*, 2002). The soybeans, developed by Monsanto from genetically modified soy plants which were resistant to the herbicide *Round-Up*, were followed by other applications aimed at commercialising the products of gene technology applied to food production, in particular commodity crops such as maize.

Various environmental non-governmental organisations (NGOs) mobilised protests against the introduction of genetically modified foods and crops, with a primary feature being the inclusion of disturbing imagery. As a consequence, media focused on these protests, and the issue of genetically modified foods attracted high levels of media attention, much of it negative. There is some evidence that this caused a (temporary) increase in consumer perceptions of risk associated with genetically modified foods, although perceptions of benefits appeared to be depressed for a longer period after the peak of media reporting (Frewer *et al.*, 2002).

One of the consequences was the adoption of stricter regulation within the European Union regarding the tracking and tracing of products containing GM ingredients, or indeed GM whole foods (EC, 2003). As public negativity towards the process of genetic modification crystallised, the introduction of novel products into the European marketplace was further compromised.

Initially, qualitative research tried to clarify what risks, and to some extent, benefits, were relevant to European consumers. The literature identified consumer concerns related to the potential for unintended effects, such as the introduction of allergies resulting from the introduction of novel or unexpected proteins in foods, horizontal gene transfer or other environmental effects, to issues directly attributable to uncertainty and unintended effects on human health and the environment, and the potential irreversibility of any negative impact. Subsequently, research grounded in both qualitative and quantitative methodology indicated that perceptions of 'unnaturalness', 'tampering with nature', animal welfare, the power balance between producers and consumers, democracy, and disparity between the industrialised world and the third world may also play a part in determining consumer responses, although the same factors may not be equally relevant to all consumers (Bredahl, 1999; Grunert *et al.*, 2001; Miles and Frewer, 2001).

The introduction of genetically modified foods into Europe illustrates many of the concepts raised in this chapter, and provides a case to show that public attitudes are not dependent on an analytical assessment of risk and benefit *per se*. Scientific research has identified the following:

1) *Psychometric dimensions*: There was little recognition across industry and government that consumers' risk perceptions were negatively biased from the outset, which could have been predicted given the *technological, unnatural* and *unknown* potential of GM foods, all factors likely to trigger negative consumer perceptions, particularly in the food domain where values and traditions are very important (Groves, 2001). Research also demonstrated that the *voluntary choice* to control consumption of GM foods was of great importance to European consumers, necessitating the labelling of GM foods and implementation of effective traceability systems, but this was not implemented in the mid-1990s (Miles *et al.*, 2005).

2) *Ethics and values*: The basis of communication about genetic modification was grounded in the 'objective' concept of 'substantial equivalence' (see for example, FAO/WHO, 2006[2000]) which built on the premise that the content of GM foods was not substantially different to conventional counterparts. However, to the public other psychological factors such as *ethical and moral* considerations, and *values* such as *concern about the integrity of nature* played a part in societal and consumer acceptance (Jensen and Sandoe, 2002; Miles and Frewer, 2001). In addition, environmental groups and environmentally aware consumers indicated that they needed more information about the potentially negative long term effects of non-contained introduction of GMO's on the integrity of the environment. The public perception that institutions and industries were pushing the introduction of genetically modified foods in order to protect their own vested interests rather than to support societal benefits did little to alleviate societal concerns (Frewer *et al.*, 2004). Evidence for this motive on the part of those with responsibility for consumer protection was provided by communication practices grounded in the principle of substantial equivalence, which was perceived at best as irrelevant to consumer concerns, and at worst as attempting to 'hide the truth' about the risks and uncertainties associated with genetically modified foods and crops.

3) *Lifestyle hazards*: Consumers had no direct experience with genetically modified foods and ingredients, and so the food products were not perceived as lifestyle hazards, but rather as technological risks. Consumers did not have positive experiences associated with the new products which in general did not provide direct and tangible consumer benefits. Taken together with low perceived controllability over exposure, a direct result of failure to trace and label

genetically modified foodstuffs, meant that optimistic bias about the risk did not develop.

4) *Systematic information processing*: When government agencies and industry relied on 'logical' arguments focusing on substantial equivalence, they (implicitly) assumed the willingness and motivation of the public either to accept their persuasion attempts (which was unlikely due to the low level of trust, and the failure to address other psychological factors), or to follow their scientific reasoning and process the arguments systematically, to arrive at the same conclusion as the experts. This was, however, unlikely due to the complex and opaque arguments used.

5) *Heuristics, trust*: At the time of the introduction of GM foods into Europe, there is evidence to suggest that government agencies in some European countries were not trusted by the general public, a particular consequence of various food crises such as the BSE-nvCJD crisis (Berg, 2004) or dioxin contamination of the food chain (Vebeke, 2001). It could, therefore, have been expected that communications by these same agencies directed towards persuading the public of the merits of the GM foods did not convince the public, particularly in the absence of broader discussion about other, potentially more negative effects. In addition, opaque risk governance systems did not develop societal trust in the activities of risk regulators, independent of the information that these same regulatory institutions supplied. Distrust in the government, together with the tendency to perceive new technology as 'risky' (Slovic, 1987) may have produced a negative affect (emotion) implicitly associated with genetically modified foods. Again, the absence of first generation products with tangible and desirable consumer benefits did little to offset this negative affect and reassure consumers about the motives of the food industry in introducing these crops and foods. Medical applications of genetic modification, with clear end-user benefit, are, for example, generally considered more acceptable by the public than those applied to food or agricultural production (Frewer *et al.*, 1997b).

6) *Emotions*: The protest campaigns by environmental lobby groups such as Greenpeace and Friends of the Earth successfully linked emotional cues to genetically modified foods exacerbating fear and disgust already in existence (Huffman *et al.*, 2004). The images of mutant tomato creatures and other *Frankenfood*, could easily grab the public attention. Moreover, these emotionally laden images in themselves communicated that the products were frightening and

disgusting, reinforcing the message the environmental groups wanted to communicate; i.e. avoiding these products was a justified act of self-protection.

The approach by government and industry illustrates that institutions, and other relevant actors in the food chain, did not realise the importance of consumer psychology in communicating about new products. The information provided by the industry and government did not address consumer concerns, and did not align with, (we assume), their preferred mode of psychological processing of persuasive information by the public. Thus much of the information provided by the proponents of GMO's was simply not relevant to consumers, or even worse, perceived to be promoting a vested interest on the part of industry and government. In contrast, the issue of visual imagery and information which aligned with the concerns of consumers, as presented by NGOs, was perceived to be more relevant and more trusted by consumers.

Future research aims

The example of genetic modification illustrates the consequences of a lack of understanding of how the public reacts to the development of new technologies, as well as commercialisation of their products. Research is urgently needed to further our understanding of the fundamental mechanisms which determine individual responses to existing and emerging food issues, particularly under circumstances where habit, emotion and information processing heuristics may have an effect on consumer decision-making. Indeed, the traditional emphasis on risk may be less relevant to consumer decision-making, as it has become increasingly evident that consumers are making decisions about the acceptability or otherwise of specific foods and production technologies based on a complex interaction of perceptions of risk and benefit associated with specific food choices.

Recent theoretical advances in the area of social psychology are relevant to the development of effective risk-benefit communication strategies. Although trust has been extensively evaluated in this context (e.g. Frewer *et al.*, 1998; Frewer *et al.*, 2003; Eiser *et al.*, 2002; Cvetkovich *et al.*, 2002; Poortinga and Pidgeon, 2005), we have reviewed some evidence to suggest that other heuristics may also be potentially influential determinants of consumer behaviour. These may include habit (Fischer *et al.*, 2006; Verplanken and Orbell, 2003) and affect (or emotion) (Slovic *et al.*, 2004). Furthermore, the role of implicit memory may generate attitudinal associations which determine whether or not information results in attitude change (Spence and Townsend, 2006). In some situations attitude activation (through inclusion of relevant cues in information) may be a more influential determinant of risk related behaviour than providing additional formal knowledge about risk and safety (Fischer *et al.*, submitted).

The relative importance of these different factors in determining attitude change, and their potential for interaction, are not well understood, and may vary across potential hazard type, indicating the need for development of case studies focusing on specific hazard types. In any case, it is inappropriate to assume that all consumers are homogenous with respect to their perceptions (whether related to trust or information needs), necessitating exploration of individual differences in this context. In particular, consumers may be differentially motivated to search for information regarding risks and indeed benefits of emerging technologies (Kornelis *et al.*, in press; Fischer *et al.*, 2006). Demographic and psychological factors may account for profound differences between different consumers regarding their responses to emerging technologies and their applications, as well as other risk issues. Systematic analysis of these is therefore required. In effect, targeted information provision is to be developed which meets the needs of different groups of consumers, as peoples' responses to risk-benefit information may also vary according to predictable individual differences (Fischer and Frewer, submitted).

Conclusions

Understanding consumer psychology is essential if we are to understand and predict peoples' responses to emerging technologies and their applications. By testing theoretical approaches derived from social psychology in specific hazard domains (for example, food risk), it is possible to generate generic hypotheses about other factors which determine technology acceptance to other domains.

In particular, we have argued that heuristics or other less conscious psychological constructs such as optimistic biases, habit, and affect are likely to play a dominant role in how consumers respond to food-related hazards, including those which may be associated with emerging technologies. One consequence is that in the introduction of new technologies, the role of heuristics and intuition should be taken into account when developing information about the risks, benefits and other salient features of the technology.

Notes

1. Two European food labels have been introduced to reflect traditional production methods and the region where a specific product has been produced. These are P.D.O. *Protected Denomination of Origin* and P.G.I. *Protected Geographical Indication*. There is also an EU label for food produced through application of organic production methods.

References

A.S. Alhakami and P. Slovic, 'A psychological study of the inverse relationship between perceived risk and perceived benefit', *Risk Analysis*, 14 (1994) 1085–1096.

L. Berg, 'Trust in food in the age of mad cow disease: a comparative study of consumers' evaluation of food safety in Belgium, Britain and Norway', *Appetite*, 42 (2004) 21–32.

L. Bredahl, 'Consumers' cognitions with regard to genetically modified foods: Results of a qualitative study in four countries', *Appetite*, 33 (1999) 343–362.

L. Bredahl, 'Determinants of Consumer Attitudes and Purchase Intentions with regard to Genetically Modified Food – Results of a Cross-National Survey', *Journal of Consumer Policy*, 24 (2001) 23.

C.M. Bruhn, 'Consumer attitudes and market response to irradiated food', *Journal Of Food Protection*, 58 (1995) 175–181.

J.T. Cacioppo, R.E. Petty, C.F. Kao and R. Rodriguez, 'Central and peripheral routes to persuasion: an individual difference perspective', *Journal of Personality and Social Psychology*, 51 (1986) 1032–1043.

S. Chen and S. Chaiken, 'The heuristic-systematic model in its broader context', in Chaiken, S. and Trope, Y. (eds) *Dual process theories in social psychology* (New York, NY: Guilford Press, 1999).

G. Cvetkovich, M. Siegrist, R. Murray and S. Tragesser, 'New information and social trust: asymmetry and perseverance of attributions about hazard managers', *Risk Analysis*, 22 (2002) 359–367.

EC (2003) Regulation (EC) No 1830/2003: http://europa.eu.int/eur-lex/pri/en/oj/dat/2003/1_268/1_26820031018eb00240028.pdf

J.R. Eiser, S. Miles and L.J. Frewer, 'Trust, perceived risk, and attitudes toward food technologies', *Journal of Applied Social Psychology*, 32 (2002) 2423–2433.

D. Evans, *Emotion: The science of sentiment* (Oxford: Oxford University Press, 2002).

FAO/WHO 'Safety aspects of genetically modified foods of plant origin: Report of a Joint FAO/WHO Expert Consultation on Foods Derived from Biotechnology'; Derived 13 November 2006 from: http://www.fao.org/es/esn/gm/biotec-e.htm

R.H. Fazio and T. Towles-Schwein, 'The MODE model of Attitude-Behavior Processes', in Chaiken, S. and Trope, Y. (eds) *Dual-Process Theories in Social Psychology* (New York, NY: Guilford Press, 1999).

C. Fife-Schaw and G. Rowe, 'Extending the application of the psychometric approach for assessing public perceptions of food risks: some methodological considerations', *Journal of Risk Research*, 3 (2000) 167–179.

M.L. Finucane, A.S. Alhakami, P. Slovic and S.M. Johnson, 'The affect heuristic in judgments of risks and benefits', *Journal of Behavioral Decision Making*, 13 (2000) 1–17.

A.R.H. Fischer, A.E.I. De Jong, E.D. Van Asselt, R. De Jonge, L.J. Frewer and M.J. Nauta (submitted), 'Food Safety in the Domestic Environment: an inter-disciplinary investigation of microbial hazards during food preparation'.

A.R.H. Fischer and P.W. De Vries (submitted), 'Everyday behaviour and every-day risk: an exploration how people respond to frequently encountered risks'.

A.R.H. Fischer and L.J. Frewer (submitted), 'Food safety practices in the domestic kitchen; Demographic, Personality and Experiential Determinants'.

A.R.H. Fischer, L.J. Frewer and M.J. Nauta, 'Towards improving Food Safety in the Domestic Environment: a multi-item Rasch scale for the measurement of the safety efficacy of domestic food handling practices', *Risk Analysis*, 26 (2006) 1323–1338.

B. Fischhoff, P. Slovic and S. Lichtenstein, 'How safe is safe enough? A psycho-metric study of attitudes towards technological risks and benefits', *Policy Sciences*, 9 (1978) 127–152.

L.J. Frewer, 'Societal issues and public attitudes towards genetically modified foods', *Trends in Food Science and Technology*, 14 (2003) 319–332.

L.J. Frewer, C. Howard, D. Hedderley and R. Shepherd, 'What determines trust in information about food-related risks? Underlying psychological constructs', *Risk Analysis*, 16 (1996) 473–486.

L.J. Frewer, C. Howard, D. Hedderley and R. Shepherd, 'The elaboration likeli-hood model and communication about food risks', *Risk Analysis*, 17 (1997a) 759–770.

L.J. Frewer, C. Howard and R. Shepherd, 'Public Concerns in the United Kingdom about General and Specific Applications of Genetic Engineering: Risk, Benefit, and Ethics', *Science, Technology & Human Values*, 22 (1997b) 98–124.

L.J. Frewer, C. Howard and R. Shepherd, 'The influence of initial attitudes on responses to communication about genetic engineering in food production', *Agriculture and human values*, 15 (1998) 15–30.

L.J. Frewer, J. Lassen, B. Kettlitz, J. Scholderer, V. Beekman and K.G. Berdalf, 'Societal aspects of genetically modified foods', *Food and Chemical Toxicology*, 42 (2004) 1181–1193.

L.J. Frewer, S. Miles and R. Marsh, 'The media and genetically modified foods: evidence in support of social amplification of risk', *Risk Analysis*, 22 (2002) 701–711.

L.J. Frewer and B. Salter, 'Public attitudes, scientific advice and the politics of regulatory policy: the case of BSE', *Science and Public Policy*, 29 (2002) 137–145.

L.J. Frewer and B. Salter, 'The changing governance of biotechnology: the pol-itics of public trust in the agri-food sector', *Applied Biotechnology, Food Science and Policy*, 1 (2003) 199–211.

L.J. Frewer, J. Scholderer and L. Bredahl, 'Communicating about the Risks and Benefits of Genetically Modified Foods: the Mediating Role of Trust', *Risk Analysis*, 23 (2003) 1117–1133.

L.J. Frewer, R. Shepherd and P. Sparks, 'The interrelationship between perceived knowledge, control and risk associated with a range of food-related hazards targeted at the individual, other people and society', *Journal of Food Safety*, 14 (1994) 19–40.

A.M. Groves, 'Authentic British food products: a review of consumer perceptions', *Journal of Consumer Studies and Home Economics*, 25 (2001) 246–254.

K.G. Grunert, L. Lahteenmaki, N.A. Nielsen, J.B. Poulsen, O. Ueland and A.N. Astrom, 'Consumer perceptions of food products involving genetic modification: Results from a qualitative study in four Nordic countries', *Food Quality and Preference*, 12 (2001) 527–542.

J. Hansen, L. Holm, L.J. Frewer, P. Robinson and P. Sandoe, 'Beyond the knowledge deficit: recent research into lay and expert attitudes to food risks', *Appetite*, 41 (2003) 111–121.

J.R. Houghton, E. Van Kleef, L.J. Frewer, G. Chryssochoidis, S. Korzen-Bohr, T. Krystallis, J. Lassen, U. Pfenning and G. Rowe, 'The Quality of Food Risk Management in Europe: Perspectives and Priorities', *Journal of Food Policy* (in press).

W.E. Huffman, M. Rousu, J.F. Shogren and A. Tegene, 'Consumer's Resistance to Genetically Modified Foods: The Role of Information in an Uncertain Environment', *Journal of Agricultural & Food Industrial Organization*, 2, (2004) article 8.

K. Jensen and P. Sandoe, 'Food safety and ethics: The interplay between science and Values', *Journal of Agricultural and Environmental Ethics*, 15 (2002) 245–253.

M. Kornelis, J. De Jonge, L.J. Frewer and H. Dagevos, 'Classifying consumers on their intended use of food safety information sources', *Risk Analysis* (in press).

C.F. Kreijl, A.G.A.C. Knapp, M.C.M. Busch, A.H. Havelaar, P.G.N. Kramers, D. Kromhout, F.X.R. Van Leeuwen, H.M.J.A. Van Leent-Loenen, M.C. Ocke and H. Verkley, *Ons eten gemeten. Gezonde voeding en veilig voedsel in Nederland* (Bilthoven, Nl, RIVM, 2004).

H.A. Kuiper, H.P.J.M. Noteborn, E.J. Kok and G.A. Kleter, 'Safety aspects of novel Foods', *Food Research International*, 35 (2002) 267–271.

J. Lassen, A. Allansdottir, M. Liakoupulos, A. Olsson and A.T. Mortensen, 'Testing times: the reception of round-up ready Soya in Europe', in Bauer, M. and Gaskell, G. (eds) *Biotechnology – The Making of a Global Controversy* (Cambridge: Cambridge University Press, 2002).

J.K. Lazo, J. Kinnell and A. Fisher, 'Expert and Lay person Perceptions of Ecosystem Risk'. *Risk Analysis*, 20 (2000) 179–193.

J.S. Lerner and D. Keltner, 'Beyond valence: Toward a model of emotion-specific influences on judgement and choice', *Cognition and Emotion*, 14 (2000) 473–493.

J.S. Lerner and D. Keltner, 'Fear, anger, and risk', *Journal of Personality and Social Psychology*, 81 (2001) 146–159.

W.J. McGuire, 'Attitudes and attitude change', in Lindzey, G. and Aronson, E. (eds) *The handbook of social psychology*, 3rd edn (New York, NY: Random House, 1985).

A.L. Meijnders, C.J.H. Midden and H.A.M. Wilke, 'Role of negative emotion in communication about CO_2 risks', *Risk Analysis*, 21 (2001) 955–966.

C.K. Mertz, P. Slovic and I.F.H. Purchase, 'Judgments of chemical risks: Comparisons among senior managers, toxicologists, and the public', *Risk Analysis*, 18 (1998) 391–404.

S. Miles, M. Brennan, S. Kuznesof, M. Ness, C. Ritsonand and L.J. Frewer, 'Public worry about specific food safety issues', *British Food Journal*, 106 (2004) 9–22.

S. Miles and L.J. Frewer, 'Investigating specific concerns about different food Hazards', *Food Quality and Preference*, 12 (2001) 47–61.

S. Miles, O. Ueland and L.J. Frewer, 'Public attitudes towards genetically modified food and its regulation: the impact of traceability information', *British Food Journal*, 107, 4 (2005) 246–262.

M.J. Nauta, R. De Jonge, A.R.H. Fischer, E.D. Van Asselt and L.J. Frewer, (in preparation, RIVM and Wageningen University), 'Towards Safer Food Preparation in the Domestic Environment: the effect of an information campaign on microbial contamination of domestically prepared food' [paper available on request from Fischer or Frewer].

K. Oatley and P.N. Johnson-Laird, 'Towards a cognitive theory of emotions', *Cognition and Emotion*, 1 (1987) 29–50.

S.M. Parry, S. Miles, A. Tridente and S.R. Palmer, 'Differences in Perception of Risk Between People Who Have and Have Not Experienced *Salmonella* Food Poisoning', *Risk Analysis*, 24 (2004) 289–299.

R.E. Petty and D.T. Wegener, 'The elaboration likelihood model: Current status and Controversies', in Chaiken, S. and Trope, Y. (eds) *Dual process theories in social psychology* (New York, NY: Guilford Press, 1999).

W. Poortinga and N.F. Pidgeon, 'Trust in risk regulation: Cause or consequence of the acceptability of GM food?', *Risk Analysis*, 25 (2005) 199–209.

G. Rowe and L.J. Frewer, 'Public participation methods: A framework for evaluation', *Science Technology and Human Values*, 25 (2000) 3–29.

P. Rozin and A.E. Fallon, 'A Perspective on Disgust', *Psychological Review*, 94 (1987) January, 23–41.

P. Rozin and T.A. Vollmecke, 'Food likes and dislikes', *Annual Review of Nutrition*, 6 (1986) 433–456.

R.A.C. Ruiter, C. Abraham and G. Cok, 'Scary warnings and rational precautions: a review of the psychology of fear appeals', *Psychology and Health*, 16 (2001) 613–630.

M. Siegrist, 'The influence of trust and perceptions of risks and benefits on the acceptance of gene technology', *Risk Analysis*, 20 (2000) 195–204.

P. Slovic, 'Perception of risk', *Science*, 236 (1987) 280–285.

P. Slovic, 'Perceived risk, trust, and democracy', *Risk Analysis*, 13 (1993) 675–682.

P. Slovic, 'Trust, Emotion, Sex, Politics, and Science: Surveying the Risk-Assessment Battlefield', *Risk Analysis*, 19 (1999) 689–701.

P. Slovic, M.L. Finucane, E. Peters and D.G. Macgregor, 'Risk as Analysis and Risk as Feelings: Some Thoughts about Affect, Reason, Risk, and Rationality', *Risk Analysis*, 24 (2004) 311–322.

P. Slovic, J.H. Flynn and M. Layman, 'Perceived Risk, Trust, and the Politics of Nuclear Waste', *Science*, 254 (1991) 1603–1607.

P. Slovic, T. Malmfors, D. Krewski, C.K. Mertz, N. Neil and I.F.H. Purchase, 'Intuitive toxicology. II. Expert and lay judgments of chemical risks in Canada', *Risk Analysis*, 15 (1995) 661–675.

B.D. Solomon and A. Banerjee, 'A global survey of hydrogen energy research, development and policy', *Energy Policy*, 34 (2006) 781.

A. Spence and E. Tonsend, 'Implicit attitudes towards genetically modified (GM) foods: A comparison of context-free and context-dependent evaluations', *Appetite*, 46 (2006) 67–74.

H. Tuorila, L. Lahteenmaki, L. Pohjalainen and L. Lotti, 'Food neophobia among the Finns and related responses to familiar and unfamiliar foods', *Food Quality And Preference*, 12 (2001) 29–37.

W. Van den Hoogen, A.L. Meijnders and C.H.J. Midden, 'De invloed van onderscheidbaarheid en saillantie van contextuele informatie op richting en sterkte van context effecten', in Holland, R., Ouwerkerk, J., Van Laar, C., Ruiter, R. and Ham, J. (eds) *Jaarboek Sociale Psychologie 2005* (Groningen, Nl, ASPO pers, 2006).

M.C. Van Putten, L.J. Frewer, L.J.W. Gilissen, B. Gremmen, A.A.C.M. Peijnenburg and H.J. Wichers, 'Novel foods and food allergies: A review of the issues, *Trends in Food Science & Technology*, 17 (2006) 289–299.

W. van Rijswijk, L.J. Frewer, D. Menozzi and G. Fiaoli, 'Consumer perceptions of traceability: a cross national comparison of associated benefits and the links with quality and safety' (in preparation).

W. Vebeke, 'Beliefs, attitude and behaviour towards fresh meat revisited after the Belgian dioxin crisis', *Food Quality and Preference*, 12 (2001) 489–498.

B. Verplanken and S. Orbell, 'Reflections on past behavior: A self-report index of habit strength', *Journal of Applied Social Psychology*, 33 (2003) 1313–1330.

N.D. Weinstein, 'Optimistic biases about personal risks', *Science*, 246 (1989) 1232–1233.

K. Witte and M. Allen, 'A meta-analysis of fear appeals: Implications for effective public health campaigns', *Health Education and Behavior*, 27 (2000) 591–615.

5
Making Sense of Uncertainty and Precaution: the Example of Mobile Telecommunications[1]

Julie Barnett and Lada Timotijevic

Introduction

The aims of this chapter are two-fold. Firstly, it aims to provide an introduction to research on public understandings of uncertainty and precaution. Secondly, it presents an analysis of focus group data collected in the context of the precautionary stance of the UK government to the possible health risks of mobile telecommunications. We note that there is often a lack of correspondence between uncertainty and concern. People are familiar with, and often accepting of the notion of uncertain science. Uncertainties that are inferred from seemingly unresponsive institutional structures are more disconcerting.

Traditionally risk is distinct from uncertainty as risk involves knowledge of both likelihood and consequences whereas uncertainty refers to situations, 'where there is no sufficient basis for assigning a precise and accurate likelihood to a particular outcome' (POST, 2004: 2). In recent years there has been an increasing requirement for transparency in risk management (House of Lords, 2000) and the uncertainties inherent in risk analysis have increasingly become the subject of scrutiny (Stirling, 2004; Frewer, Miles, Brennan *et al.*, 2002). The government responses to the Bovine Spongiform Encephalopathy (BSE) inquiry (HM Government, 2001) noted 'the need to be open about uncertainty and to make the level of uncertainty clear when communicating with the public'. Explicit recognition and transparent communication of the range of uncertainties often associated with scientific and technological developments is thus increasingly required of policy-makers and politicians. Over the last ten years one area in which these issues have played out is around the management of the potential health risks associated with mobile telecommunications (MT) (Green Alliance, 2001). In 2000 the Independent Expert

Group on Mobile Phones (IEGMP) report acknowledged that the possibility of unwanted health risks occurring as a function of radiofrequency radiation could not be ruled out. In the wake of the recommendations the report contained, the UK government adopted a precautionary approach to managing the potential health risks. A range of precautionary actions were implemented and advice provided in public information leaflets (Department of Health, 2000a and b; see too NRPB, 2004).

This is the backdrop against which we will explore public understandings of both uncertainty and precaution. Following a brief introduction to previous literature around public understandings of uncertainty and precaution we will go on present some qualitative work that explores public understandings of, and responses to, uncertainty and of the precautionary approach in relation to MT in the UK.

Why mobile telecommunications?

The last decade has seen huge growth in mobile phone use: recent figures show that between 2000 and 2004 the total number of minutes of mobile rose from 34 to 62 billion. There are now 61.2 million mobile phone subscribers in the UK – a number greater than the UK population – 85 per cent of UK households have mobile phones and 27 per cent of all calls are made from mobile phones (OFCOM, 2005). For the first time in 2005 the proportion of households exclusively relying on mobile phones (10 per cent) is the same as the proportion who only use landline phones (OFCOM, 2006). It is in the context of this remarkable uptake of the technology that there has been ongoing expert attention, and an intermittent media and public focus, on the possible association between radio waves and negative health impacts.

The IEGMP investigated the scientific basis for linking mobile phone signals with negative health impacts on concentration, memory and attention as well as cancers and effects on the cardiovascular, endocrine and immune systems. In the ensuing report (IEGMP, 2000) uncertainties were explicitly recognised as it concluded that,

> ... it is not possible at present to say that exposure to RF radiation, even at levels below national guidelines, is totally without potential adverse health effects, ... the gaps in knowledge are sufficient to justify a precautionary approach. *(para 6.35–6.42)*

Examination of the policy discourse reveals that the commissioning of the IEGMP report was in part constructed as a response to public concerns (for full details of this see Timotijevic and Barnett, 2006). The precautionary actions and advice that the report recommended were considered to be a way of addressing these concerns. For example, in accepting this approach, the government response to the IEGMP report was explicit in anticipating the effect of a precautionary approach upon public concern (Department of Health, 2002): 'The report makes helpful recommendations on measures to reduce public concern about the health impacts of MT technologies' (para 1.2).

Public understandings of uncertainty

In considering the way in which publics make sense of uncertainty and precaution around MT it can be noted that the health risks associated with mobile phones are generally seen as rather less serious than a range of other risks (Poortinga and Pidgeon, 2003). The benefits of mobile phones are highly salient (Petts, Wheeley and Homan *et al.*, 2003) and over recent years there has been a trend towards fewer people believing that handsets are bad for health (MORI, 2004).

There is a growing body of research around public perceptions of uncertainty. The first theme of this work is around the impact of uncertainty upon perceptions of risk or the source of the risk communication. Although it seems that communication of risk can itself cause increased concern (Morgan, Slovic and Nair *et al.*, 1985; McGregor, Slovic and Morgan, 1994), it is not necessarily the case that communication of uncertainty produces greater concern than 'certain risk' estimates do (Johnson and Slovic, 1995; Kuhn, 2000; Bord and O'Connor, 1992). There is some evidence suggesting that uncertainty can reduce motivation to act as it may confuse, lead to complacency and be used to discount the seriousness of the threat (Roth, Morgan, Fischhoff *et al.*, 1990; Maule, 2004). On the other hand, under some circumstances the communication of uncertainty may increase the credibility of the information source (Johnson and Slovic, 1995, 1998). A substantial body of qualitative work suggests that it is denials of uncertainty and claims of safety that are more likely to be mistrusted than admissions of uncertainty (Wynne, 1992; Grove-White, Macnaughton, Mayer *et al.*, 1997). In contrast recent work by Frewer, Hunt, Kuznesof *et al.* (2003) in relation to food risks suggests that experts often believe that publics expect absolute rather than uncertain estimates of risk.

The second theme of research on public perceptions of uncertainty focuses on factors that may affect the way in which people make sense of uncertain public health information. Differences in value systems (Kuhn, 2000) and varying formats for presenting risk information (Roth, Morgan, Fischhoff *et al.*, 1990) may both affect responses to uncertain information. The nature of the risk itself also seems to be important. Miles and Frewer (2003) found that the communication of uncertainty increased the perception of risk for hazards that were under societal rather than personal control.

Public understandings of precautionary action and advice

Earlier we noted the way in which the precautionary approach to managing the uncertainties around the health effects of mobile telecommunications in the UK was in part instrumentally constructed as a way of reducing public concerns. Since the IEGMP report, a small body of literature in Europe has addressed the extent to which precautionary approaches *do* have the effect of reducing public concerns.

The work of Burgess (2004) provided an early impetus to research in this area. Drawing upon a mast action group case study and business and market research data, he argued that precautionary actions and advice signal the existence of risk which in turn triggers protest activity and intensified media presentations of risk. An essay by Sandman (2004) suggested that on balance precaution was more likely to raise concerns than to reassure.

Similar conclusions were reached in two studies using experimental designs. Wiedemann, Thalmann and Grutsch *et al.* (2006) largely replicate the conclusions of Weidemann and Schütz (2005) in concluding, 'that precautionary measures may trigger concerns and amplify EMF-related risk perceptions' and that 'information about the implementation of precautionary measures has no positive effect on trust in public health protection' (p. 361).

Within the UK, a large-scale survey (n=1742) explored awareness of the precautionary advice contained within the DoH leaflet about mobile phone health risks, and public responses to it (Barnett, Timotijevic and Shepherd *et al.*, in press). In line with previous research, precautionary advice was generally associated with increased concern rather than providing reassurance. In particular, those who reported higher levels of concern about the uncertainties associated with MT risk – that is those ostensibly in most need of reassurance – were less reassured by precautionary advice than those with lower concern about uncertainty. The authors however noted the way in which the nature of the measurement tool itself may have affected these findings, as the closed-ended response options strip the interpretative framework of real life context, provide little opportunity for expressing shades of meaning and may encourage respondents to read risk from uncertainty. Indeed, as will be showed below, qualitative work has suggested that causing concern or providing reassurance are not the only ways in which people evaluate the provision of precautionary advice (Timotijevic and Barnett, 2006).

In short, massive expansion of mobile telecommunications, the recognition of the uncertainties about the health risks associated with the MT and the emergence of precautionary approaches to regulating the possible MT health risks represent a useful backdrop against which to explore the way in which people make sense of and respond to uncertain risks and the associated precautionary approach. In the remainder of this chapter we will report some qualitative work that aimed to explore these issues.

A note on methods

A series of nine focus groups were conducted in London and Brighton. Full details of the composition of the groups, the sampling strategy and the way that group discussions were structured can be found in Timotijevic and Barnett (2006). Suffice it here to say that participants recruited in a London borough characterised by a high profile media debate and public protests about the siting of a mobile phone base station were defined in relation to their stated level of concern and whether they had taken part in protests. Participants in Brighton were grouped by age and by whether they were parenting young children. This strategy was adopted in order to access a broad range of views of uncertainty and precaution around both mobile phones and base stations.[2]

Qualitative analysis software (NVivo) was used to develop a coding system within groups. Converging and diverging views were then identified by comparing and contrasting the resulting themes across the groups.

Risks and benefits of MT: setting the scene

It is helpful to situate our consideration of the way in which people make sense of uncertainties around possible health risks associated with MT in the context of perceived risks and benefits of mobile phone use. The early stages of the focus groups contained general discussion of the role of mobile phone use in the lives of the group participants. For many of them, the benefits of mobile phone use were considerably more salient than any possible adverse health effects associated with phones and masts. Mobile phones were seen to be an indispensable part of modern life – even to those that were concerned about possible health risks.

> *You get trapped into it. I didn't even want a mobile phone. I'm not technologically astute [...] My husband bought me one, he made me have one. [...] I've been forced into using it because everybody else uses one and it's the only way that you can get in touch with someone because everyone leads such busy lives [...]. (London, Protest Group, Female).*

Even in the context of strong appreciation of the value of phones, the association of MT with health risks seemed quite feasible, if not inevitable. This in itself was not solely due to the shortcomings of science or those managing the risk but rather was extrapolated from the view that a great deal of modern life and the technologies that define many everyday activities carry some level of uncertainty with them. The perceived benefits of mobile phones, their ubiquitous nature and wide penetration were some of the other warrants used in discounting the possible risks and concerns associated with mobile phones.

> *It is concerning a little bit, I must admit, but not enough to stop me making phone calls on it. In actual fact, I am probably the world's worst culprit because I have a desk phone next to me and, because all the numbers are logged in my mobile [...] I can't bothered to look up the phone number and then dial it from the fixed phone, instead I call from the mobile. (Brighton, Age 30–50, Male)*

> *[My teenage daughter] has her phone, she uses it, the damage is done if it's going to be done. (Brighton, Age 30–50, Female)*

Negative health effects tended to be more strongly associated with masts than with mobile phones. Arguments for the greater potential of masts for negative health effects were justified in terms of their greater size and their indiscriminate effects.

> *[Masts] are probably like a giant mobile phone but worse. (London, Concerned, Not Active group, Male)*

Masts can affect younger children who haven't got mobile phones. If you've chosen to have a mobile phone then you take the risks involved and you can use it as often or not, as you want. [...] (London Protest group, Female)

In contrast, the physical experience of mobile phone use (e.g. the heating of the head when in use) as well as the numbers using phones were arguments invoked to support a claim that handsets pose greater risks than masts.

Mobile phones are more concerning because there are more people affected. (Brighton, Age 30–50, Male)

I get the impression you get more harm holding a phone up to your ears than standing 200 metre away from a mast. (Brighton, Age 18–30, Male)

I worry about how much time I spend on it. I think, oh, my head's getting warm, I must get off now. (London, Not Concerned, Not Active, Female)

Such inferences did not go unchallenged, as this exchange illustrates:

M1   *When my ear's getting hotter, I change ears. And that is genuinely true. After a long phone call, maybe 20 minutes, it will feel very, very warm.*
M2   *What does it tell you? It's an indication?*
M1   *That my ear is hot.*
F   *It uses energy. Anything that uses energy gets hot. It doesn't mean it's doing anything to you. (Brighton focus group, Age 30–50)*

In considering the possible risks from MT, participants commonly drew upon the context afforded by other risk domains. They drew comparisons with issues such as smoking and asbestos where early conclusions about their negative impacts on public health had been substantially revised in the light of later scientific advances. Thus, for example:

While I agree with our responsibility, the problem is that often the scientists don't even know the answer. For example, 50 years ago doctors were saying smoking was fine. We now know that smoking is not fine. Fifty years ago, electricity pylons were fine and now we know they're not fine. I don't want to take the risk, a wait-and-see approach, I understand that we can't have guarantees but if there is a risk, I don't want to take it. (London, Non-Protest Parents, Male)

Risk comparisons were also used to minimise the validity of concern about MT:

M   *But that's happening everywhere. You've got huge cables in the ground giving off radiation.*

F1   *Sitting in front of the computer for any length of time is supposed to be dangerous as well.*

F2   *Microwaves are as well.*

F3   *They say never stand in front of it.*

M   *It's where you draw the line because if you think about it all the time, you wouldn't get up. Because I mean, people don't really understand the amount of things that are actually going on all the way around them. (Brighton, Age 50+)*

Making sense of uncertainty

Analysis of the focus group transcripts identified two main ways in which participants made sense of uncertainty. First, acknowledged uncertainty – where people recognised uncertainty in scientific estimates of the health risks of MT technology. Second, inferred uncertainty – another level of uncertainty occurs when people infer uncertainty either from expert conflict or from institutional arrangements that appear impermeable to the possibility of acknowledging uncertainty. These perspectives on uncertainty were warranted with reference to personal experience, the actions and advice of government and industry, both around MT itself and other risks.

It is important to note at this stage that uncertainty is not necessarily associated with concern. Where there was a widespread appreciation of the benefits of mobile phone use this was either linked with contending that the risks were small or with discounting of more substantial uncertainties.

It's bad for you but there's no proof and you can't trust the man who's from the company to tell you the full story. He won't say much anyway but unless they come up with some hard evidence it's not going to concern me. (London, Not Concerned, Not Active, Male)

Acknowledged uncertainty

Many people showed a clear appreciation and acceptance that estimates of the risk of negative health effects associated with MT were uncertain. To some extent this was grounded in an appreciation of the nature of scientific investigation and the factors that constrain the conclusions that can be drawn.

> M   *'I think it's rather unproven what damage they do, so it's rather an unknown quantity.*
>
> *[...]*
>
> F1   *Well, that's the most worrying thing – that we don't know.*
>
> *[...]*
>
> F2   *The trouble is that it probably won't be proven for many years because the damage is probably quite slow working. (London, Concerned, Not Active)*

> *It's very difficult to measure [...] With thalidomide you, sadly, had physical examples, and the one connecting strain, if you like, of those people was that all their mothers had taken thalidomide. It's more difficult to say my son or daughter's got leukaemia because we live near a phone mast. (London, Protest Group, Male).*

There were two particularly salient dimensions of acknowledged uncertainty: firstly, about who will be harmed and secondly, about when the harm will occur. Other dimensions of uncertainty such as what the harm involved seemed to be less important. These findings echo the work of Frewer, Miles and Brennan (2002) who found that not only do people recognise different dimensions of uncertainty but also that some dimensions are more likely to evoke concern than others.

Who will be harmed?

The uncertainties about possible health risks of mobile phones were in part articulated by focusing on the question of who would be harmed. It was suggested that some people may be more susceptible to the potential risks of MT than others. In part this argument noted the likely range of variation across individuals. In addition, however, it was argued that certain people (such as children, older people and those that are unwell) may be particularly vulnerable.

> *Some people may be particularly susceptible to those particular lengths of waves, radio waves, whatever it is that are coming off them. But on average I wouldn't think most people suffer that much. (Brighton, Age 50+, Female)*

The mast really affects innocents, younger children, who have got nothing to do with mobile phones, or the elderly, who might not be mobile phone users. It's just indiscriminate, really. (London Protest group, Female)

You think about your children and their health risks because the longer term effects are not known. (London, Non-Protest Parents, Female)

The uncertainty around who might be harmed, coalescing as it did around an inequitable distribution of the risks, and the particular vulnerability of children, led to expressions of concern for many.

When might the harm occur?

Another set of uncertainties were acknowledged around when possible harms from MT might occur. It was considered quite possible that there was no scientific evidence of health effects now but that such effects may emerge in years to come. In the situations where evidence might be delayed and cumulative, some believed that experts could not be conclusive.

The trouble is that it [health risk] won't be proven for many years because the damage is probably slow at working. (Brighton, Age 50+, Male)

F1 *There's no smoke without fire so maybe there is something in years to come to find out exactly what the danger is.*
F2 *It's impossible to assess long-term impact. There's not been enough time. (London, Non-Protest Parents)*

For some, this uncertainty around the time lag between exposure to the technology and the emergence of scientific evidence about the existence (or absence) of health risks constituted a cause of concern. Here, the conditions of scientific uncertainty were equated to a form of live experiment; the publics were seen as 'guinea pigs' in testing the technology.

F1 *It's a classic case of us not having lived long enough to see what develops later on in life with the children who've grown up in the area. It's like a lot of other things. We don't know now what's happening.*

F2   *We're really guinea pigs.*
F1   *It'll only be in 20 or 30 years time that they'll be able to say, yes, this caused this. (London, Protest Group)*

For others, uncertainties about when health effects might occur offered the possibility of warranting a lack of concern. Thus:

It doesn't bother me. In 20 years time I'm sure they will have invented something to get rid of a tumour if I find one on the side of my head from my phone.

Inferred uncertainty

Whether or not acknowledged, the existence of uncertainties may also be inferred. We identified two circumstances in which this occurred. Firstly, uncertainty was inferred from conflicting *expert* estimates of risk. Secondly, uncertainty was inferred, even when experts apparently agreed, should there seem to be a lack of clear and transparent communication. In both instances, described below, inferring uncertainty provided a way of actively making sense of the situation.

Expert conflict

The first situation in which uncertainty was inferred was where, as is commonly the case, sources of risk communication provide different and conflicting estimates of the risk (Breakwell and Barnett, 2002). This may provide a challenge that can be resolved in various ways. Awareness of conflicting risk estimates is not necessarily associated with concern.

Some participants saw little evidence of expert conflict around MT, rather considering that experts were generally in agreement that there were a few risks:

...There are scientific groups that study this in a number of different countries and there's a major group based in the Netherlands, it's an international organisation group, and they study the issue as well and they have not arrived at the conclusion that it is harmful. So, to some degree, you have to accept expert point of view. (London, Non-Protest Parents, Female)

For others, risk communication seemed to be characterised by multiple positions on whether MT was linked with negative effects on health.

I think the information we get is conflicting. For every expert that says it's safe, there's another who says it's not safe. (Female, London, Non-Protest Parents)

This conflict was sometimes linked with concern. In the example below a comparison with the controversy around the vaccination for mumps, measles and rubella (MMR) was deployed to depict the concerns stemming from conflicts between experts:

Well it's a bit like the row about the MMR jab. You've got all the doctors saying it's safe, you've got another doctor saying it might be dangerous. Unless you're a scientist you're never going to know. I think round here, unless anybody's a doctor, how will we know? You just have to take those things in the newspapers and try and make your own mind up and worry about your own kids. (Brighton, Age 50+, Male)

On other occasions expert conflict was linked with uncertainty but not with concern. The excerpt below represents conflict as healthy and necessary for the personal process of reaching an informed judgement.

F   *The best way is always to have a discussion when you've got the pro and the anti together because if some are for it, you only hear the one side and you do need somebody countering their arguments and for them to be able to counter back so I always think that's the best way. People can always put forward a very plausible argument for their own case but you always need somebody there to pick the holes and in it and for them to be able to reply, so whichever way it's done, I always think you have to have the for and against.*

M   *It allows you to make a judgement based on both sides of the case. But nobody expects them to be certainties. (London, Non-Protest Parents)*

Apparent expert certainty

The second situation in which uncertainty was inferred occurred where there was general expert agreement that there was a small likelihood of negative impacts. If this position was seen to be taken by those regulating the risks (in conjunction with what were considered to be inappropriate institutional liaisons and a lack of clear and honest communication) the inference was drawn that rather than small risks, there were considerable uncertainties. In the focus groups it was clear that many participants felt that there were discrepancies between what they knew and what they believed to be the information available about MT health risk. In order to substantiate their contention that there was much more uncertainty than was admitted, people drew on two main arguments around the actions and practices of stakeholders in MT.

Firstly, inferences about scientific uncertainties were drawn in the context of experiences around mast siting practices and planning permission procedures. There was believed to be a lack of clear and timely information about planning applications for masts on community sites and inappropriately close relationships between councillors and operators. One person commented:

> *It's almost as if they try and come in overnight, erect these masts, and then just go away. And then you find out about them when it's too late. So, if there isn't anything wrong with them, why is it done in such an underhand way?'. (London, Protest Group, Male)*

Secondly, uncertainty was more likely to be inferred in situations where the presenting face of risk regulators was seen to be unresponsive and impenetrable. It is hard for people to see how the structures underlying such an interface can be sensitive to, and able to admit new information. One way this played out was in relation to government and industry. The knowledge of the high profits generated by expanding MT applications and the benefits accruing to government and industry, signalled to many focus group participants that the government would be unable or unwilling to be motivated to regulate the technology with appropriate care.

> *It's not in anybody's interest actually to tell anybody that phones are dangerous if they are. Because the government is making money out of them, as are the phone companies. (Brighton, Age 18–30, Male)*

> *It is in the interests of the mobile phone companies because the more conflicting information there is, the more uncertainty people have and the less they can make informed opinions. It could be that we are all panicking totally unnecessarily and that there isn't anything wrong, but we won't know that. (London, Protest group, Male)*

It can be noted here that it was in this context that a strong preference for 'independent' research was expressed. Publics are sometimes considered naïve in expressing such a wish and for not realising that 'no-one is really independent'. However for the focus groups participants, the desire for independent research stems from concerns about the ostensibly impenetrable and unresponsive structures that they sense hedge about the regulation and management of MT risks. In this context independent scientific research is seen to have the potential capacity to cut through this layer and has the potential to access uncertainties. Research by mobile phone companies will, by definition, not have this capacity. One participant noted:

> *I think everyone would welcome an independent research provided it was genuinely that – independent, not muddied by anyone with an axe to grind or a profit to make on it, and they were informing a willing public about the dangers that we don't know enough about. I think we're all just burying our heads in the sand a little bit because it's very convenient ... If we were made aware .... (Brighton, Age 30–50, Male)*

In the following excerpt, the protests about phone masts are explicitly constructed as a reaction to unresponsive structures of governance.

> *Frankly, it doesn't matter who we chuck in or out of the town hall in [name of the council] because they're pretty much powerless to act. They don't damn well listen to us. The objection to mobile phone masts was that we have no democratic voice and that we are bound to be stroppy about it. (London, Non-Protest Parents, Male)*

We had little evidence in the focus groups that inferences of uncertainty that were based upon unresponsive structures of risk managers led to increased concern about the possible association of MT technology and negative health effects *per se*. Indeed in line with the qualitative work noted above, in the situation where expert categorisations of small, apparently known, risks are apparently not open to challenge, people may express a *preference* for such acknowledged uncertainties. Inferences of uncertainty were rather linked with the lack of trust in risk regulators and generally high levels of scepticism and even cynicism about the risk management approaches in the UK.

Uncertainty and precaution

The focus groups illustrated that on occasions both inferred and acknowledged uncertainties could be used to warrant the expression of a lack of concern and to justify continuing high levels of phone use and a lack of precautionary action.

> *The ambiguity gives people an excuse not to worry about it because it's not definite. It's not been agreed by everyone that using them is bad. When it is, people will say that using it is bad and they'll think of ways around it, whatever. But, while there's that kind of is it or isn't it, people will just carry on in blind ignorance basically. (London, Not Concerned and Not Active, Male)*

On other occasions, where uncertainty and concern *are* aligned alongside each other, the value of a more precautionary stance in managing these uncertainties was suggested:

> F1   *I don't think there's any proof that it's not harmful.*
> F2   *I don't think any of them can really say it is or it isn't.*
> F3   *So that leaves an element of doubt and when there's doubt there's a problem.*
> F2   *I'd think you'd rather not have it there though, if you don't know.*
> F4   *It's no good afterwards them saying sorry. (London, Protest Group)*

This allows us to briefly consider the question of the role that precautionary actions and advice may have in relation to concern. As well as exploring understandings of uncertainty, the focus groups reported above were also used to explore the ways in which people made sense of precautionary advice around mobile phones.[3]

Although, as seen from the above excerpts, group participants recognised the existence of scientific uncertainty, it was rarely considered that this uncertainty had been acted upon by the government and that the government had adopted a precautionary stance. The following exchange also indicates this:

> Interviewer: *Do you know if the government has a position on the issues related to health risks of mobile telecommunications?*
> F   *No.*
> *[General agreement]*
> M   *Probably wait and see.*
> (Brighton, 18–30 age group)

There was a considerable discrepancy between what people believed to constitute a credible precautionary approach and the nature of the current approach. Participants drew on a range of evidence in order to substantiate the stance that it was not credible that the government took a precautionary approach. The improbability of a precautionary approach was often warranted with references to the relationship between government and industry and a consequent scepticism as to the motives of the regulator.

They have to have set rules and regulations but at the end of the day they're not going to start blurting out that if you use this phone it's going to do this, this and this to you because they'd lose their money.
(London, Not Concerned and Not Active Group, Female Respondent)

There was no evidence in the groups that learning about the existence and nature of the precautionary action/advice *initiated* concern. Rather, in line with the literature reviewed above, concern about the precautionary approach was expressed by those who had already articulated concern about the uncertainties associated with MT. In this instance people used precaution to confirm their concerns about emissions from masts and their dissatisfaction with those managing the risks. Here a precautionary approach was used to warrant existing scepticism, as suggested by the following discussion:

IV   *How does it make you feel that they have adopted ... this strategy?*
F1   *It makes me think that we're right, then, with our concerns. If there were no concerns they would be presenting the evidence.*
F2   *We probably wouldn't be sitting here either.*
M1   *It's quite a harsh statement as well. It implies that something is actually wrong.*
M2   *But then it doesn't sit, does it? So, once again they say we won't do that, and all the figures are massaged. I do take your point about it being all governments, but I wouldn't believe anything this government says. If they said something was black I'd immediately believe it was white.* (London, Protest group)

Irrespective of whether or not there was concern about the possible health risks of MT, precautionary advice and actions were often considered to confirm the perceived inability of institutions to manage risks. They were seen as a way of enabling government to 'cover their backs' and avoid accountability in the event of any eventuality. There was a sense in which raising the spectre of risk and uncertainty with serial risk communications was meaningless in the context of everyday life.

It was considered that the motivation of such a pattern of communication was simply to enable the justification of 'we told you so' at a later date should possibilities later become realities. For example:

M   *They're trying to cover themselves for later on.*
F   *But that's just their response to everything, isn't it? Yes and maybe, perhaps, depends on what happens. We'll see. We don't know.*
    (Brighton, 18–30 age group).

For others, a precautionary stance did offer some reassurance. This was not necessarily reassurance about the risks from MT being minimised but rather that it provided evidence that the government could be responsive to uncertainty.

M   *I don't know. I mean, if they don't know, they don't know. So they can't say it's fine to use them because they're acknowledging there's a potential.*
F   *At least they're acknowledging that there's a potential so I don't think it's that negligent...* (London, Not Concerned, Not Active Group).

Conclusions

We have seen that people made sense of uncertainties not only from what they knew about the science of MT risks, but by drawing upon the wider context in which risk communication and management took place. There was little evidence of overlap between an acknowledgement of scientific uncertainties and concern. In contrast where uncertainties were inferred from unresponsive institutional practices this often led to expressions of concern about what might *not* be being acknowledged by those regulating the risks. Similarly, the nature of current precautionary actions and advice around MT were often considered inadequate insofar as they left unchallenged what were seen to be often inappropriate relationships between government and industry. When communications drawing attention to possible risks came from – or were associated with – seemingly unresponsive institutional structures, they were often seen to signal a lack of accountability rather than, as might be intended, as being indicative of an open and transparent approach to risk management.

This analysis illustrates the situated nature of risk perceptions (as outlined especially in Chapter 3 and Horlick-Jones, 2005). It also resonates with the work of Weidemann *et al.* (2003) who draw attention to the way in which understandings of risk are driven by shared interpretative patterns of the social context in which the risk emerges. It is also instructive to consider the work of Bates *et al.* (2005). They demonstrated in relation to genetic research, the importance of the ways in which people warrant their views and in doing so draw on shared experiences. This suggests the importance of institutions responding to concerns in ways that take account of, and correspond to the warrants that different groups employ to support their concerns.

The results reported here clearly resonate with work around other technologies, and as indicated by other chapters in this volume. For example in a study exploring public understandings of genetically modified foods, Marris, Wynne and Simmons (2001: 10) noted how participants 'did not ask for zero risk or full certainty with respect to the impact of GMOs and were well aware that daily activities of ordinary lives have to be balanced against one another'. They noted too that it was the institutional denial of uncertainty not its existence that was most problematic. This also links with the work of Frewer, Miles and Brennan *et al.* (2002) who found uncertainties related to the knowledge limitations of science to be more acceptable than those stemming from government regulatory activity – or lack of it.

At one level we might conclude that a carefully devised approach to communicating scientific uncertainty is needed that takes account of the views of different groups and the ways in which they are warranted. The findings reported here however suggest that the effectiveness of such an approach may be blunted by the fact that it is the uncertainties inferred from unresponsive institutional structures that are most deep rooted and persistent. We have also questioned the value of precautionary actions and advice in reassuring publics about scientific uncertainties. In this regard we note the approach of Stirling (2002, 2004) where the focus of precaution shifts to 'more long term, holistic, integrated and inclusive social processes for the governance of risk than are typically embodied in conventional risk assessment' (Stirling, 2002: 22). Here the question of whether precaution heightens or reduces public concern is largely rendered redundant. The focus rather shifts to an approach to regulating uncertainty, ambiguity and ignorance around science and technology that actively engages the public. Arguably such an approach addresses the root of the response to uncertainties inferred from the perceived lack of institutional responsiveness. The focus here is not on the assumed relationship between transparency and trust but rather on enabling appropriate and critical involvement of the public when developing policies around uncertain science.

Notes

1. The work this chapter is based on was undertaken by Julie Barnett, Lada Timotijevic, Richard Shepherd, Vicky Senior and Jane Vincent (University of Surrey) who received funding from the Mobile Telecommunications and Health Research Programme. The views expressed in the publication are those of the authors and not necessarily of the funding institutions.
2. Base stations were commonly referred to by focus groups as 'masts'. This terminology will thus be used from now on in this chapter.
3. These data are fully reported in Timotijevic and Barnett (2006).

References

J. Barnett, L. Timotijevic, R. Shepherd and V. Senior, 'Public responses to precautionary information from the department of health (UK) about possible health risks from mobile phones', *Health Policy* (in press).

B.R. Bates, J.A. Lynch, J.L. Bevan and C.M. Condit, 'Warranted concerns, warranted outlooks: a focus group study of public understandings of genetic research', *Social Science and Medicine*, 60 (2005) 331–344.

R.J. Bord and R.E. O'Connor, 'Determinants of risk perceptions of a hazardous waste site', *Risk Analysis*, 12, 3 (1992) 411–416.

G. Breakwell and J. Barnett, 'The significance of uncertainty and conflict: developing a social psychological theory of risk communications', *New Review of Social Psychology*, 2, 1 (2002) 107–114.

A. Burgess, *Cellular phones, public fears and a culture of precaution* (Cambridge: Cambridge University Press, 2004).

Department of Health, *Mobile phones and health* (London: Department of Health, 2000a).

Department of Health, *Mobile phone base stations and health* (London: Department of Health, 2000b).

Department of Health, *Mobile Phones and Health: Government responses to the report from the IEGMP (Stewart Group)* (London: Department of Health, 2002).

http://www.dh.gov.uk/PublicationsAndStatistics/Publications/PublicationsPolic yAndGuidance/PublicationsPAmpGBrowsableDocument/fs/en?CONTENT_ID =4096744&MULTIPAGE_ID=4903699&chk=EM/9E%2B   Accessed October 2006.

L. Frewer, S. Hunt, S. Kuznesof, M. Brennan, M. Ness and C. Ritson, 'Views of scientific experts on how the public conceptualise uncertainty', *Journal of Risk Research*, 6, 1 (2003) 75–85.

L. Frewer, S. Miles, M. Brennan, S. Kuznesof, M. Ness and C. Ritson, 'Public preferences for informed choice under conditions for risk uncertainty', *Public Understanding of Science*, 11 (2002) 363–372.

Green Alliance, *Decision making under scientific uncertainty: the case of mobile phones* (Green Alliance, 2001: available at http://www.green-alliance.org.uk/ publications/PubMobilePhones/ Accessed October 2006).

R. Grove-White, P. Macnaghton, S. Mayer and B. Wynne, *Uncertain World: Genetically Modified Organisms, Food and Public Attitudes in Britain*

(Lancaster: Centre for the Study of Environment Change, University of Lancaster, 1997).

HM Government, *The Government Response to the BSE Enquiry* (London: The Stationery Office, 2001).

T. Horlick-Jones, 'Informal logics of risk: Contingency and modes of practical reasoning', *Journal of Risk Research*, 8, 3 (2005) 253–272.

House of Lords, *Science and Society, Select Committee on Science and Technology, Third Report, HL 38, Feb 2000* (London: HMSO, 2000).

IEGMP, *Mobile phones and Health* (Didcot, UK: NRPB, 2000).

B.B. Johnson and P. Slovic, 'Explaining uncertainty in health risk assessment: Initial studies of its effect on risk perception and trust', *Risk Analysis*, 15 (1995) 485–494.

B.B. Johnson and P. Slovic, 'Lay views on uncertainty in environmental health risk assessment', *Journal of Risk Research*, 1, 4 (1998) 261–269.

K.M. Kuhn, 'Message format and audience values: Interactive effects of uncertainty information and environmental attitudes on perceived risk', *Journal of Environmental Psychology*, 20 (2000) 41–51.

C. Marris, B. Wynne, P. Simmons and S. Weldon, *Public Perceptions of Agricultural Biotechnologies in Europe*. Final Report of the PABE research project funded by the Commission of European Communities Contract number: FAIR CT98-3844 (DG12 – SSMI) (2001). Accessed 02/01/07 from http://www.lancs.ac.uk/depts/iepp/pabe/docs/pabe_finalreport.pdf

A.J. Maule, 'Translating Risk Management Knowledge: The lessons to be learned from Research on the Perception and Communication of Risk', *Risk Management: an International Journal*, 6, 2 (2004) 15–27.

D.G. McGregor, P. Slovic and G.M. Morgan, 'Perception of Risks from Electromagnetic Fields: A Psychometric Evaluation of a Risk-Communication Approach', *Risk Analysis*, 14, 5 (1994) 815–828.

S. Miles and L.J. Frewer, 'Public perception of scientific uncertainty in relation to food hazards', *Journal of Risk Research*, 6, 3 (2003) 267–283.

M.G. Morgan, P. Slovic, I. Nair, D. Geisler, D. Macgregor, B. Fischhoff, D. Lincoln and K. Florig, 'Powerline Frequency Electric and Magnetic Fields: A Pilot Study of Risk Perception', *Risk Analysis*, 5, 2 (1985) 139–149.

MORI, *Mobile Telephony and Health: Public Perceptions in Great Britain*. Research Study conducted for GSM Association, Mobile Manufacturers Forum and Mobile Operators Association (London: MORI, Feb 2004).

NRPB, *Mobile Phones and Health 2004*, Report by the Board of National Radiological Protection Board, Documents of the NRPB, Vol. 15 No. 5 (2004). Last accessed 02/01/07 from http:www.hpa.org.uk/radiation/publications/documents_of_nrpb/pdfs/doc_15_5.pdf

OFCOM (2005), The Communications Market http://www.ofcom.org.uk/research/cm/cm05/ Accessed October 06.

OFCOM (2006), The Communications Market http://www.ofcom.org.uk/research/cm/cm06/ Accessed October 06.

J. Petts, J. Wheeley, J. Homan and S. Niemayer, *Risk Literacy and the Public: MMR, Air Pollution and Mobile Phones* (London: Department of Health, 2003).

W. Poortinga and N. Pidgeon, 'Exploring the Dimensionality of Trust in Risk Regulation', *Risk Analysis*, 23 (2003) 961–972.

POST (2004) Handling uncertainty in scientific advice, Postnote, June 2004, No. 220. Available at http://www.parliament.uk/documents/upload/ POSTpn220.pdf Accessed October 2006.

E. Roth, M.G. Morgan, B. Fischhoff and L. Lave, 'What do we know about making risk comparisons?', *Risk Analysis*, 10 (1990) 375–387.

P. Sandman, 'Because people are concerned: How should public "outrage" affect application of the precautionary principle' (2004). Available at http://www.psandman.com/articles/vodafone.pdf Accessed October 2006.

A. Stirling, 'Risk Uncertainty and Precaution: Some instrumental implications from the social sciences', in Scoones, I., Leach, M. and Berkhout, F. (eds) *Negotiating change: Perspectives in Environmental Social Science* (London: Edward Elgar, 2002).

A. Stirling, 'Opening up or Closing Down? Analysis, participation and power in the social appraisal of technology', in Leach, M., Scoones, I. and Wynne, B. (eds) *Science Citizenship and Globalisation* (London: Zed, 2004).

L. Timotijevic and J. Barnett, 'Managing the Possible Health Risks of Mobile Telecommunications: Public Understandings of Precautionary Action and Advice', *Health, Risk and Society*, 8, 2 (2006) 143–164.

P.M. Weidemann, M. Clauberg and H. Schutz, 'Understanding amplification of complex risk issues: the risk story model applied to the EMF case', in Pidgeon, N., Kasperson, R.E. and Slovic, P. (eds) *The Social Amplification of Risk* (Cambridge: Cambridge University Press, 2003).

P.M. Wiedemann and H. Schütz, 'The precautionary principle and risk perception: Experimental studies in the EMF area', *Environmental Health Perspectives*, 113 (2005) 402–405.

P.M. Wiedemann, A.T. Thalmann, M.A. Grutsch and H. Schütz, 'The Impacts of Precautionary Measures and the Disclosure of Scientific Uncertainty on EMF Risk Perception and Trust', *Journal of Risk Research*, 9, 4 (2006) 361–362.

B. Wynne, 'Uncertainty and Environmental Learning: Preconceiving science and policy in the environmental paradigm', *Global Environmental Change*, (1992) June, 111–127.

6
Against the Stream: Moving Public Engagement on Nanotechnologies Upstream

Alison Mohr

Introduction

Nanotechnologies are materials and devices conceived, developed and applied on a scale of one thousand millionth of a metre. Nanoscale materials differ not only in size but also in behaviour to their larger-scale counterparts. Nanomaterials have a greater surface area relative to material produced on a larger scale that potentially renders them more chemically reactive, thereby altering certain molecular properties. Matter at the nanoscale is also susceptible to behavioural changes dominated by quantum effects, which may affect the optical, electrical and magnetic behaviour of nanomaterials. Nanotechnologies are intrinsically multidisciplinary; amalgamations of materials science, chemistry, physics, engineering, life and medical sciences.

The infinite potential of nanotechnologies to change the technological, economic as well as social landscapes for future generations has recently prompted exponential government funding of research in the United Kingdom and Europe to harness its benefits. Yet, while the spectre of the genetic modification (GM) controversy continues to produce social disquiet, public concerns regarding nanotechnologies' unintended social and environmental effects and impacts are emerging. In light of recent GM debates, how might publics satisfactorily and meaningfully engage with innovation processes surrounding nanotechnologies to ensure their safe and ethical development and application?

In the public debates surrounding GM, issues of social concern were narrowly framed as 'impacts' or 'risks', restricting their consideration 'downstream' of the innovation process, after the technology had been stabilised or 'black-boxed'. Science and Technology Studies

(STS) approaches have demonstrated that black-boxes can be opened up to public scrutiny as technologies are a heterogeneous mix of technical and social relations that cannot be separated (Law and Hassard, 1999). It can be shown that innovation processes often involve 'upstream' assumptions about the social adoption of a new technology, how it will be used, by whom, and for what purpose. This could be achieved by building in opportunities for public engagement 'upstream' of the innovation process before it closes and becomes black (Latour, 1997).

This chapter sets out to critically discuss the notion and value of upstream public engagement in the context of recent academic, policy and public debates on nanotechnologies and with particular reference to the Royal Society/RAEng report's call for engagement processes to be built in to the innovation process.

The emerging public debate

Grey goo, cyborgs and the amplification of weapons of mass destruction are some of the potential hazards of nanotechnology flagged up by the vociferous Canadian action group on Erosion, Technology and Concentration (ETC) when it called for a moratorium on the commercialisation of nanotechnology in February 2003. In the same month, the UK Government's Better Regulation Taskforce (2003) called for the development of a new regulatory framework for nanotechnology and for an early and informed dialogue between scientists and the general public about its impacts; to which the UK Government (2003) replied there was no obvious focus for an informed public debate at the time but that it would keep the position under review.

Nevertheless, the issue of nanotechnologies undoubtedly entered the public sphere when Prince Charles' erroneously reported fears about self-replicating nanomachines capable of smothering the world in grey goo made international headlines. The ensuing public controversy prompted the UK Government to commission The Royal Society and The Royal Academy of Engineering (RAEng) to investigate the potential opportunities and uncertainties posed by nanotechnology. Prince Charles further ignited the nanotechnology debate when in July 2004 he published an article in the *Independent on Sunday* warning of the potential dangers of nanotechnology. While the Prince acknowledges that nanotechnology is a triumph of human ingenuity he calls on those promoting the technology to show 'significantly greater social awareness, humility and openness' than they did over GM (Prince of Wales, 2004). He warns the public will accept nanotechnology only 'if a precautionary approach is seen to be applied', urging 'regulatory processes' to be 'encouraged to develop at the same rate as the technology' and for risk assessment to 'keep pace with commercial development'.

Indeed, research and development on nanotechnologies is progressing exponentially. In 2005, expenditure worldwide rose to more than US$9.5 billion (RNCOS, 2006). The vast majority of which was funded by various governments and corporate institutions. By 2008, it is estimated that the global market for nanotechnologies will exceed $28 billion. In 2003, the UK Government pledged £45 million per annum from 2003 to 2009 as part of its nanotechnology strategy (House of Commons, 2004). Meanwhile the European Commission has increased its expenditure on nanotechnologies from €1.3 billion in the Sixth Framework Programme (2002–06) to €3.5 billion in the forthcoming Seventh Framework Programme (2007–13) (CORDIS, 2006).

In response to increasing concerns regarding current and future developments of nanotechnologies expressed by non-governmental organisations (NGOs) such as ETC and by some nanotechnologists, and prompted by increased research funding by government and industry, The Royal Society and The Royal Academy of Engineering (2004) published a report entitled, *Nanoscience and Nanotechnologies: Opportunities and Uncertainties*. Authored by an unusually inclusive expert working group,[1] it considered the current and future developments of nanotechnology through consultation with a range of stakeholders as well as drawing on the group's own expertise. Wide-ranging social issues such as impacts on human health and the environment as well as the consideration of safety, ethical and societal concerns by way of stakeholder and public dialogue were considered alongside the current state of scientific knowledge and the potential future uses of nanotechnologies and their responsible development. Areas where additional regulation may be required were also investigated. The report acknowledged that nanotechnologies have the potential to provide numerous benefits to society and recommends steps to realise this potential while minimising possible future uncertainties and risks. One of those steps was the recommendation that the research councils fund a more sustained and extensive programme of research into public attitudes to nanotechnologies to draw out the wider social and ethical issues various publics wish to raise and to track any changes as public knowledge about nanotechnologies develops.

Public attitudes towards nanotechnologies

In actual fact the publics' perceptions of the potential risks and benefits of new technologies are regularly surveyed as a means for informing research and development funding programmes and public policy. In recent years, a number of studies in the UK and in Europe have examined public attitudes towards nanotechnologies to sketch an upstream picture of public attitudes towards these novel developments and in response to the cautionary tale of the GM controversy. Among these were the 2002 and 2005 Eurobarometer surveys which each contained a single question relating to nanotechnologies. Eurobarometer surveyed 16,500 citizens aged 15 and over across 15 member states in 2002 increasing to 24,895 citizens across 25 member states in 2005 (European Commission Directorate General Research, 2002, 2005). A brief survey complemented by two in-depth workshops was used to inform the Royal Society/RAEng report in 2004. Designed to give a basic measure of awareness of nanotechnologies among members of the public aged 15 or over, it comprised three questions with a representative survey of 1005 people in Great Britain. At the European level, the NanoDialogue project, funded under the European Commission's Sixth Framework Programme, surveyed 663 visitors aged 15 and over who attended exhibitions on nanotechnologies (influenced by the contents of the Royal Society/ RAEng report) staged in seven science centres and museums across Europe from March to August 2006.[2] The exhibitions aimed to provide information and raise awareness among European publics on the latest research in nanotechnologies while the related survey hoped to identify the main issues and preoccupations of these publics concerning nanotechnologies. A question common to the Eurobarometer, Royal Society/RAEng and NanoDialogue surveys was 'what effect do you think nanotechnology will have on our way of life in the next 20 years?'.

In the three years between the Eurobarometer surveys, the number of respondents who thought that nanotechnologies would improve our way of life in the next 20 years rose from 30 to 48 per cent, while those who thought it would make it worse also rose, albeit marginally, from 6 to 8 per cent. The high level of 'don't know' responses, which decreased from 53 per cent in 2002 to 40 per cent in 2005, indicates low general levels of awareness on the issue of nanotechnologies across Europe.

Public attitudes are widely recognised as playing a crucial role in realising the potential of technological advances by the Royal Society's expert working group who commissioned a survey by BMRB

(British Market Research Bureau) of public opinion which ascertained that public awareness of nanotechnologies in Great Britain is low. Only 29 per cent had heard of 'nanotechnology' while even fewer, 19 per cent, could offer a definition. Of this 19 per cent it is interesting to note that in spite of the hitherto predominantly negative media coverage, 68 per cent felt it would improve our way of life while only 4 per cent thought it would make our way of life worse. Two workshops conducted by BMRB with the public to explore their views in more depth revealed the public held positive views about new advances and potential applications including medicine, the creation of new materials, and improvements to quality of life. Their concerns focused on the financial implications of nanotechnologies and their potential impact on society. The reliability of new applications and any potential long-term side effects they may produce, as well as our ability to control them, were also of concern. The issue of governance was also deemed important; in particular, how to ensure that the development of nanotechnologies is socially beneficial. Inevitably, comparisons with GM and nuclear power were made by the participants.

The visitors to the exhibitions that formed the backbone of the NanoDialogue project were also asked to indicate their level of understanding of nanotechnologies prior to entering the exhibition. Only 15 per cent of respondents thought that their level of understanding was high while 56 per cent indicated a low level of understanding. Over a fifth of respondents, 19 per cent, claimed that they had no understanding of nanotechnologies at all. Little or no understanding of nanotechnologies again do not appear to have had a detrimental effect on respondents' views of the effect of nanotechnologies on our way of life as only 3 per cent thought it would make things worse, while 58 per cent thought that our way of life would improve. Two per cent were of the opinion that it would have no effect at all on our way of life, while a significant 37 per cent weren't sure of the effect that nanotechnologies would have.

Since 1973, the European Commission has conducted regular sampling of public opinion in member states to support its policies as well as to evaluate their success. These tools are used not only as policy support instruments but also to map the evolution of European opinion as the Commission strives to construct a united European polity. However, inferences regarding the European publics' opinions and attitudes reported in the Eurobarometer results have drawn widespread criticism for drawing a distinction between 'facts' and 'values' which previous research on risk has rendered problematic or indefensible (Wynne, 2001). The Eurobarometer treats risk and trust as independent qualities assuming that risk is a quantitative measure of probability while trust is a qualitative measure of personal opinion and therefore value-based. Critics of this distinction argue that qualitative social scientific research has demonstrated that close ties exist between perceptions of risk and historically and culturally conditioned expectations about the trustworthiness of governing institutions (Irwin and Wynne, 1996). Risk and trust are therefore interdependent and not independent variables. Indeed, the responses elicited by public opinion surveys, Eurobarometer or otherwise, tend to vary widely according to what the study is designed to reveal. Consequently, publics and their understandings are often poorly portrayed by the studies designed to characterise them (Campbell and Townsend, 2003).[3]

In spite of the methodological and consequent epistemological flaws inherent in such instruments, Eurobarometer and its kind have given rise to an active debate about the characteristics of the citizen in relation to technoscientific issues. It is constantly debated among academic and policy circles whether citizens or wider publics are ignorant or informed, fearful or confident, drawn in by or resistant to national cultural and political framings (Jasanoff, 2005).

A new mood for public engagement?

The low general levels of awareness of nanotechnologies coupled with the publics' positive outlook of their potential effect on our way of life, alluded to by these surveys, suggests that the governance of nanotechnologies would be an appropriate subject for initial public dialogue, particularly given that the UK government is currently funding social scientific research into nanotechnologies.

Detecting a new mood for public constructive, inclusive and open public debate and dialogue, the government's latest ten-year strategy for science and innovation incorporates a commitment to:

> enable the debate to take place 'upstream' in the scientific and technological development process, and not 'downstream' where technologies are waiting to be exploited but may be held back by public scepticism brought about through poor engagement and dialogue on issues of concern. (HM Treasury, 2004: 105)

The Royal Society/ RAEng report recognises that 'many of the issues currently surrounding nanotechnologies are "upstream" in nature' and calls for 'a constructive and proactive debate about the future of the technology now, before deeply entrenched or polarised positions appear. Our research into public attitudes highlighted questions around the governance of nanotechnologies as an appropriate area for early dialogue' (The Royal Society and The Royal Academy of Engineering, 2004: 66–67).

In response to the Royal Society/RAEng's report, the UK Government (HM Government, 2005) asserts its continuing commitment to initiating public dialogue, which will inform both the direction of research and development and the progression of appropriate regulation. To this end, in August 2005 the Government published its Programme for Public Engagement on Nanotechnologies (HM Government, 2005). In response to the Royal Society/RAEng recommendation 'that the government initiates adequately funded public dialogue around the development of nanotechnologies' (The Royal Society and The Royal Academy of Engineering, 2005: 67), the government reiterated its 'wish to make substantial and sustained progress towards building a society that is confident about the governance, regulation and use of science and technology ... [and] to ensure that debate takes place at an early stage' (HM Government, 2005: 1). At the core of the public engagement programme are two projects which existed prior to the Royal Society/RAEng report, both modestly funded in 2004 by Sciencewise, the Office of Science and Technology's (OST) grant scheme. However, only one of these projects, Nanodialogues (which builds upon the valuable research previously funded by the Economic and Social Research Council (ESRC) of Lancaster University and Demos), engages directly with the public. Nanodialogues use a variety of upstream engagement methods involving stakeholder, representative and randomly selected participants. These methods investigate the practicalities of public engagement for identifying risk and informing regulation, for shaping research goals and identifying its potential uses in the corporate innovation cycle, and in identifying opportunities, barriers and benefits to the global diffusion of nanotechnologies.

The other public engagement project, the Nanotechnologies Engagement Group (NEG), is a network of research organisations whose role is to map the current practice of public engagement on nanotechnologies and to measure the understanding of stakeholders' expectations of such activities. The NEG will 'ensure that the Government's programme builds upon best practice in public engagement, supports the development of that practice and ensures that public engagement feeds into policy and decision-making' (HM Government 2005: 4). In doing so, it aims to develop a deep understanding of the principles and practices of upstream public engagement by reflecting on the lessons learned in relation to nanotechnologies in a strategic review of the development of upstream engagement, due in 2007.

A third project, even more modestly funded by Copus,[4] is 'Small Talk'. 'Small Talk' aims to bring coherence to a series of activities around the UK focused on discussions between publics and scientists on issues relating to nanotechnologies. The publics' and scientists' attitudes which emerge from these events, including some staged during the BA Festival of Science, aim to inform the research agendas of the Research Councils and the regulatory frameworks of the government. The only new Government-funded initiative mentioned in the Programme for Public Engagement is a series of stakeholder meetings designed to inform the decisions of the Government's Nanotechnologies Research Coordination Group (NRCG). The programme content is made to look more substantial through the inclusion of four non Government-funded UK projects and three European projects.

The Government's establishment of a Voluntary Reporting Scheme for industry and research organisations to identify potential risks and the conduct of a comprehensive review of the adequacy of current regulatory controls is to be applauded (Department of Trade and Industry, 2006). However, its lack of commitment to establishing an ongoing, robust programme of public engagement on nanotechnologies and its failure to propose additional funds to research their future social impact, in spite of recognising that nanotechnologies pose potential risks, was a disappointing blow to the Royal Academies (The Royal Society, 2005).

The troubled relationship between science and society

In order to present a critique of upstream public engagement's potential, its limitations as well as its misinterpretations, it is useful to discuss the notion within the wider context of the troubled relationship between science and society. Rip (2005) describes the traditional relationship between science and society as one of 'impactors' and 'impactees'. According to Rip, this dichotomy signifies a difference in power but is more significantly a difference in timing. Impactors of technological developments are active in determining the trajectory of a new technology at an early stage while the impactees, the genuine spokespersons for society and the consumers of the developed technology, have little choice but to take delivery of the end product. The difference in timing that results in society being presented with a technological *fait accompli* is discussed by Collingridge (1980) who identified the (scientific) knowledge and (public) control dilemma which this situation creates.

At an early stage of the development process, the visions and interests of the developers are still negotiable – but the downstream effects and impacts are still unclear. Once the developed technology becomes socially embedded, the vested interests of the developers, indeed the technology itself, become difficult to change. This 'dilemma' articulates a further asymmetry linked to the 'promotion' of a technology by its developers and its 'control' by users – an asymmetry apparent in the governance of science and society. On the one hand, government policy almost always focuses on the promotion of technology, emphasising its potential benefits to society. While on the other, the government's own agencies are concerned with reducing the human and social costs of new technologies through risks assessments and regulations. This dichotomy between the promotion and control of technology is part of what Rip (2005) calls the *de facto* constitution of modern industrialised societies.

An implication of this process of *de facto* constitution, whereby decisions regarding technology development are taken by a select few, is that programmes of technology assessment, public awareness and engagement will remain fragmented if they are restricted to focusing on solving (downstream) problems relating to agenda setting, legitimacy and socio-technological conflict. Technology assessment and public engagement form part of a larger co-evolutionary process and their success depends not only on how they fit into that process, but at which points in the technological trajectory and to what effect? It is not enough to simply evaluate (downstream) a range of technology assessment and public engagement exercises to identify and compare 'best practice'. As the following section demonstrates, certain ideological barriers to implementing upstream engagement persist.

Persisting paradigms of risk

Much of the debate surrounding the relationship between science and society has been narrowly defined in terms of a technical definition of 'risk' – insinuating that risk can be easily predicted, managed and controlled by expert institutions and public policy as deemed necessary. This narrow framing of risk assumes that publics either recognise or misunderstand risk in the same technical terms. Thus a key challenge for public policy is to interpret and communicate risks in realistic and practical rather than technical terms. Furthermore, issues whose interpretations by publics are widely varied or ambivalent are presumptively framed as ones of risk, as if risk could be defined universally among divergent publics. Thus a broader recognition of the idiosyncratic perception of risk is needed by policy-makers to develop socially-relevant and inclusive public policy.

The tendency to frame uncertain or even unknown technological consequences as risks is common practice in scientific practice and public policy. Thus public debates framed in terms of risk tend to focus on the *downstream* consequences of technological development. Questions about *upstream* agenda-setting – the process of innovation and about whose priorities and visions are in the 'driving seat' – are left unanswered. The general assumption is that the publics' concerns are focused on potential risks and consequences rather than on human needs, aspirations and expectations which shape tacit knowledges as well as innovation.

It is for this reason that technology assessment and public engagement activities such as consensus conferences and citizens' juries, which aim to 'open-up' (Stirling, 2005) technology innovation processes to public scrutiny and participation have been labelled the 'new tyranny' (Cooke and Kothari, 2001). The pre-framing of the issues and methods of participatory activities, the imposition on these of narrow scientific framing and of the presumptive 'normative' conceptions of the role and knowledges of publics has led to a closing-down' (Stirling, 2005) of wider innovation and policy discourses. Wider and more effective public engagement thus requires a move beyond a preoccupation with methods and procedures to embrace a more fundamental socio-political analysis of science and technology encompassing upstream concerns of agency, power, accountability and democracy. However, in trying to simply insert public engagement upstream of the innovation process, the (re-)creation of new tyrannies must be avoided, prompting some important questions asked by Jasanoff (2005). Who makes the choices that govern people's lives? On whose behalf? In which forums and with which discourses? With what right of representation? According to whose moral/ ethical code?

Persisting rhetorics of linearity

Contributing to the notion of the 'closing down' of innovation processes and the creation of new tyrannies is a persisting rhetoric of linearity that is regularly applied to innovation trajectories. This linear conception assumes the uncomplicated application of public engagement along certain points of the innovation trajectory. However, this notion precludes opening up innovation to alternative trajectories and possibilities and instead succeeds in closing it down by restricting engagement to certain points and times in the process. Wilsdon *et al.* (2005: 36) reason that 'like deficit models of the public, linear models of innovation are a default, unthinking response to the complexity of the subjects they purport to describe'. Rhetorics of linearity perpetuate myths about the inevitability and inflexibility of technological advancement. The possibility of an alternative science or a different technology goes unconsidered in the rush to fruition and possible commercialisation. Yet, innovation in practice does not advance in a line. Rather it is a process of continuous interaction and negotiation between networks of scientists, technologists, engineers, financiers, consumers and corporations.

Thus a linear model of innovation implies that the inclusion of public engagement at particular points of the innovation trajectory may alter the pace of technological advancement but not its course. This deficit model of public engagement precludes a more complex role for public engagement whereby 'scientists and engineers, sensitised through engagement to wider social imaginations, might for themselves decide to approach their science and innovation differently' (Wilsdon *et al.*, 34). The suggestion that social values and interests could themselves become the source of alternative innovation trajectories has regrettably been overlooked by the otherwise commendable Royal Society/RAEng (2004) report. In stating that public attitudes play a crucial role in the realisation of the potential of technological advances, the report lapses back into the familiar deficit model of the public by inferring that upstream forms of public engagement with science are about the earlier prediction and ensuing management of impacts (Wynne, 2006).

In criticising the linear model of innovation, we also need to turn our critical lens to the linearity of the metaphorical 'stream' of engagement. Wilsdon *et al.* (2005: 38) argue that 'a limitation of the notion of 'upstream' engagement is its implication that we can move up and down innovation processes at will, inserting a bit of public engagement where we judge it will be most effective'. By restricting public engagement to specific points in the innovation process, even upstream, we risk closing down alternative trajectories and possibilities. Rather we need to approach the metaphorical 'stream' not as an object or a named thing but as an action – of moving in a continuous flow. In this metaphorical context, upstream engagement – at a point where innovation trajectories are still malleable – is only the start of an ongoing process of deliberation and public assessment that entrenches dialogue between scientists, stakeholders and publics in all stages of the innovation process, thus allowing the possibility that the flow or direction of innovation my be altered.

Moreover, it has been suggested that we imagine upstream engagement as an ongoing cycle:

> ... it seems evident that different models of engagement are suitable at different stages. In general, where the research is in early stages and especially where it is leading-edge and complex and there is great scientific uncertainty about outcomes, benefits and risks, small scale deliberation between scientists and others will tend to be most appropriate. Once applications and consequences are more evident, either anticipated or already realised, mass participation methods become more relevant. (Jackson *et al.*, 2005: 353)

Nevertheless, there are some who argue that science is not a democratic activity amenable to public engagement. Dick Taverne, a Liberal Democrat peer and founder of 'Sense About Science' – a charity to promote the evidence-based approach to scientific issues – disputes 'the fashionable demand by a group of sociologists for more democratic science, including more "upstream" engagement of the public and its involvement in setting research priorities' (Taverne, 2004: 271). Taverne's comments were provoked by an editorial in *Nature* (2004) which argues that upstream engagement '... is worth doing – provided that all involved consider two points before beginning. First the process must be long-term and properly funded ... More importantly, funding organisations must make a genuine commitment to react to the results of engagement processes'.

In acknowledging the metaphorical, ideological as well as the practicable barriers to undertaking upstream engagement, one possible entrée into the upstream reaches of the innovation process to address socio-political concerns is that referred to by Latour (1983) in 'Give me a laboratory and I will raise the world'. Latour argues that implicit in laboratory practices and the processes of 'doing' science are a whole range of assumptions in which the relations between the inside (technical) and the outside (social) worlds are continually negotiated and translated. That is, they are never stable. In order to expose the societal dimensions of science and technology, Latour (1997: 21) encourages us to 'follow scientists ... through society' in real-time to experience the institutional and cultural contexts in which they are situated: 'instead of black boxing the technical aspects of science and then looking for social influences and biases, we realised how much simpler it is to be there before the box closes and becomes black'.

Lessons to be learned from GM

Before embarking on a journey following nanoscientists through society, it would seem prudent to take a few moments to reflect on the lessons to be learned from the public debates surrounding the GM controversy. The Royal Society/RAEng report is widely regarded as incorporating a considered response to the socio-technical controversy which enveloped GM in its attempt to anticipate and address broader societal concerns surrounding nanotechnologies. On the surface the report appears to signal a new approach to technological governance through the consideration of social and ethical impacts and stakeholder and public dialogue relating to nanotechnologies. However, the question of risk still dominates as much of the analysis and subsequent recommendations is devoted to the assessment of risk to human health and the environment. Grove-White *et al.* (2004) argue that this is not surprising given that when faced with new technologies, policy-makers tend to revert to familiar tools and ideological frames of reference. Just as risk assessment models originally developed in the nuclear industry in the 1970s shaped subsequent policy discussions surrounding GM, so discussions around nanotechnology are likely to be shaped by models devised for GM. Research into the distinctive character and properties of nanotechnology is therefore paramount to developing an adequately customised regulatory framework. Likewise, the unique and potential risks and benefits posed by nanotechnologies need to be opened up to public scrutiny and debate.

This poses an interesting question about other lessons to be learned from GM. Is there significant insight to be gained from reflecting on the GM controversy and the public debates surrounding it? Kearnes *et al.* (2006) have identified a further four lessons for nanotechnologies, based on their research of the competing understandings of 'the science' as well as of 'the public' which emerged from the GM controversy. First, the emergence of NGOs as key actors in socio-technical controversies such as GM is regarded with suspicion and ignorance by many scientific and policy actors. Research funded by the ESRC found that GM scientists' and industry's perceptions of NGOs was that of anti-GM organisations who manipulated public opinion for self-interest (Cook, 2005). On the contrary, Fischer (2003: 219) argues that NGOs' representation of traditionally marginalised public interests encourages policy debates 'grounded in local realities and citizen inter-pretations rather than would-be "objective realities" designed by ana-lysts sitting behind desks'. Thus a more in-depth understanding of the ways in which NGOs may 'represent' differentiated and multiple publics in wider society is needed.

Second, the particular conduct of the GM Nation? debate high-lighted the persistence of the deficit model of the public understanding of science. The publics' assumed scepticism or mistrust continues to pose a barrier for institutions involved in the assessment and regula-tion of new technologies. Third, the GM controversy further demon-strates the way in which the publics' concerns tend to concentrate around broader and often unpredictable points or issues. A case in point being that the British public focused its attention on more general concerns such as the possible consequences for human health and the environment. Most notably, issues of corporate ownership of patents and the power corporations wielded in the policy-making process were of particular concern. Yet these types of concerns can be symptomatic of new technologies in general and are not just the product of the GM controversy. By employing upstream engagement processes, the underlying dynamics that lead to the emergence of wider public concerns could be investigated and accommodated in the innovation process and before the technology becomes locked-in (Macnaghten *et al.*, 2005). Finally, the research conducted by Demos and Lancaster University has served to highlight the ways in which sci-entists imagine the social role of their technology to be. These scientific imaginaries – future possibilities presented by genetic modification – were not subject to public scrutiny or debate, or even tacitly acknowledged, in GM debates. However, in order to understand how such visions are performed, we need to investigate them to dis-cover what they are, how they were formed and what influence they hold over research and policy agendas.

While there are obvious lessons to be learned from the public debates surrounding GM, these should encourage (rather than deter) us to continue to pursue new methods with which to engage publics to reflect on the particular societal issues and concerns raised by nano-technologies so as to develop customised public policy and regulatory parameters. In spite of the growing literature on the inadequacies of current risk assessment paradigms, current regulatory frameworks continue to rely on the deficit model of the public understanding of science, as Chapter 2 suggests. Indeed, the enduring notion of the public understanding of science perpetually embraced by some sections of the scientific and policy communities, places at risk the upstream consideration of the wider societal implications of nanotechnologies. Even though it is widely accepted that social, ethical and moral implications of nanotechnologies should form an integral component of the upstream and continual development and assessment of nanotechnologies, the technocratic expert-led approach persists.

It is also important to note that nanotechnologies are currently at an earlier developmental stage than biotechnology when public controversy surrounding GM erupted in the late 1990s. Thus the particular technological applications and potential impacts of nanotechnology are difficult to define for public deliberation as they have yet to materialise. However it is argued that their wider social deliberation, consideration and negotiation 'upstream' of the decision-making process may benefit their social acceptability and legitimacy. Indeed, one of the criticisms levelled at the UK Government-funded *GM Nation? Public Debate* held in Britain in the summer of 2003 was that it took place too late to influence the direction of GM research or to alter the institutional, economic and political commitments of key actors. While so-called 'green (agri-food) biotechnologies' have been widely criticised for too little public intervention, 'red (medical) biotechnologies' have been subject to high levels of public participation. One reason why public debates on human biotechnology are considered to be more successful is that deliberative processes began early and kept pace with scientific developments.

Undoubtedly, the potential of nanotechnologies promises to be revolutionary. Yet the continuing GM controversy has demonstrated what can happen when the underlying social visions of key science and policy actors is not made transparent and opened up to public scrutiny and debate. The challenge to those involved in the research and development of nanotechnologies therefore is to articulate their visions, expectations and concerns surrounding the technology at an early stage by exposing it up to upstream public engagement with wider publics.

Conclusions: opening up alternative possibilities?

This chapter set out to explore how publics might satisfactorily and meaningfully engage with innovation processes surrounding nanotechnologies to ensure their safe and ethical development and application. Important lessons have been learned from the publics' ineffectual engagement with GM issues, where persistent paradigms of risk restricted debate to downstream issues such as health and environmental risks and other impacts.

Recent STS approaches have successfully demonstrated that nano-technologies, as in all technologies, are constitutive of social relations. Therefore innovation processes often involve upstream assumptions about the social adoption of nanotechnologies by envisaging its potential uses, by whom and to what effect. By building in opportunities for engagement upstream in the innovation process, publics may be able to alter the course of a nanotechnology, or even halt its development altogether.

Detecting a new mood for effectual engagement, the UK Government set out to establish its support by creating opportunities for upstream public engagement. Most notably, the Royal Society/RAEng report, commissioned by the government, has been influential in embedding the notion of, and the desire to strive for, upstream public engagement in debates surrounding nanotechnologies.

However, several overlapping challenges in the form of remnants from deficit modes of thinking about the relationship between science and society threaten to undermine the potential of upstream public engagement. First, Rip (2005) argues that the developers or promoters of a nanotechnology will have *power* disproportionate to the users who will have limited control over the nanotechnology's intended use.

Second, this imbalance of power also signifies a difference in *timing* that is articulated by Collingridge's (1980) dilemma. The dilemma being that once a nanotechnology has become socially embedded downstream of the innovation process, the visions and interests of its developers, which were previously negotiable upstream, become difficult to change.

Third, *persisting paradigms of risk* limit public engagement to the downstream consequences of nanotechnology development. Upstream issues such as who is driving the agenda, and whose visions and interests are shaping the innovation process are left unaddressed. Deficit understandings of the role and knowledges of publics in engagement activities will lead to a closing down (Stirling, 2005) of wider innovation and policy discourses on nanotechnologies. Careful attention is therefore needed to avoid the engagement tyrannies (Cook and Kothari, 2001) of the GM controversy, of simply trying to insert and address downstream concerns upstream.

Finally, contributing to this notion of the closing down of nano-technology innovation processes and the creation of new tyrannies are *persisting rhetorics of linearity* – relating to innovation and, reflexively, to the metaphorical 'stream'. A linear conception of innovation implies the uncomplicated application of public engagement to certain points and times in the nanotechnology innovation trajectory. However, this notion precludes the opening up of nanotechnology innovation to alternative trajectories and possibilities. This assumes that nano-technology innovation advances in a straight line rather than the product of convoluted interactions and negotiations between networks of actors. Likewise, a metaphorical 'stream' of engagement implies that we will be able to move up and down the nanotechnology innovation processes by applying public engagement where we judge it will be most effective. Rather what is needed is a paradigm shift to realise streams of engagement moving in a continuous flow. Thus upstream engagement is only the start of an ongoing process of deliberation and public assessment that entrenches dialogue between scientists, stake-holders and publics in all stages of the nanotechnology innovation process.

By imagining upstream public engagement as a continuous flow, and through the application of different models and scales of engage-ment to different stages of the nanotechnology innovation process, we stand the best chance of avoiding the metaphorical, ideological and practicable tyrannies characteristic of the GM polemic.

Notes

1. The expert working group combined expertise from a range of scientific and social scientific disciplines including: engineering, chemistry, nanotechno-logy, political philosophy, environmental science and environmental and occupational medicine. See http://www.nanotec.org.uk/workingGroup.htm for further details.
2. The seven science centres and museums include: Fondazione IDIS-Città della Scienza, Italy; Centre de Culture Scientifique, Technique et Industrielle de Grenoble, France; Flanders Technology International Foundation, Belgium; Deutsches Museum, Germany; Universeum AB, Sweden; Ciência Viva-Agência Nacional para a Cultura Científica e Tecnológica, Portugal; and Ahhaa Science Centre, Estonia.
3. Using the GM Nation? as a case study, research undertaken by Campbell and Townsend (2003) at Nottingham University on *Methodological Issues in Public Attitude Research*, funded by the Leverhulme Programme, found that public attitude surveys often result in biased samples resulting from self-selecting participants, leading and vague questions and the lack of contextualisation of questions.

4. The now defunct Copus Grant Schemes have, since 1987, supported and encouraged ways of making science accessible to public audiences in the UK, with funding from the Office of Science and Technology and the Royal Society.

References

Better Regulation Task Force, *Scientific Research: Innovation with Controls*. Better Regulation Task Force report (London: Cabinet Office Publications and Publicity Team, 2003).

S. Campbell and E. Townsend, 'Flaws Undermine Results of UK Biotech Debate', *Nature*, 455 (October 2003) p. 559.

D. Collingridge, *The Social Control of Technology* (London: Pinter, 1980).

G. Cook, *The Presentation of GM Crop Research to Non-Specialists: A Case Study*. Full Research Report, ESRC R000223725 (2005). Available: http://www.esrcsocietytoday.ac.uk/

B. Cooke and U. Kothari (eds), *Participation: the New Tyranny?* (London: Zed Books, 2001).

CORDIS, *Nanoscience and Nanotechnology in the EC Research Programmes* (Brussels, 2006). Available: http://cordis.europa.eu/nanotechnology/src/ec_programmes. htm

Department of Trade and Industry, *Nanotechnology 2 Year Review*, letter from Lord Sainsbury to Professor Sir John Beringer of the Council for Science and Technology (October 2006). Available: http://www.dti.gov.uk/files/file34430.pdf

European Commission Directorate General Research, 'Europeans and Biotechnology in 2002', *Special Eurobarometer 58.0*, 2nd Edition (Brussels: March, 2003). Available: http://ec.europa.eu/public_opinion/archives/ebs/ebs_ 177_ en.pdf

European Commission Directorate General Research, 'Social Values, Science and Technology', *Special Eurobarometer 225* (Brussels: June 2005). Available: http://ec.europa.eu/public_opinion/archives/ebs/ebs_225_report_en.pdf

F. Fischer, *Reframing Public Policy. Discursive Politics and Deliberative Practices* (New York: Oxford University Press, 2003).

R. Grove-White, M.B. Kearnes, P. Macnaghten, P. Miller, J. Wilsdon and B. Wynne, *Bio-to-Nano: Learning the Lessons, Interrogating the Comparison*. Project working paper presented at a high level seminar at the Royal Society (June 21 2004).

HM Government, *Response to the Royal Society and Royal Academy of Engineering Report: 'Nanoscience and Nanotechnologies: Opportunities and Uncertainties'* (February 2005). Available: http://www.dti.gov.uk/files/file14873.pdf

HM Government, *The Government's Outline Programme for Public Engagement on Nanotechnologies* (August 2005). Available: http://www.dti.gov.uk/files/file27705.pdf#search=%22upstream%20engagement%22

HM Treasury, *Science and Innovation Investment Framework 2004–2014* (July 2004). Available: http://www.hm-treasury.gov.uk/spending_review/ spend_sr04/ associated_documents/ spending_sr04_science.cfm

House of Commons, *Too little too Late? Government Investment in Nanotechnology*, House of Commons Science and Technology Committee, Fifth Report of Session 2003–04, Vol. 1 (London: 2 April 2004).

A. Irwin and B. Wynne (eds), *Misunderstanding Science? The Public Reconstruction of Science and Technology* (Cambridge: Cambridge University Press, 1996).

R. Jackson, F. Barbagello and H. Haste, 'Strengths of Public Dialogue on Science-Related Issues', *Critical Review of International Social and Political Philosophy*, 8, 3 (2005) September.

S. Jasanoff, *Designs on Nature. Science and Democracy in Europe and the United States* (Princeton: Princeton University Press, 2005).

M. Kearnes, P. Macnaghten and J. Wilsdon, *Governing at the Nanoscale. People, Policies and Emerging Technologies* (London: Demos, 2006).

B. Latour, 'Give Me a Laboratory and I Will Raise the World', in Knorr-Cetina, K.D. and Mulkay, M.J. (eds) *Science Observed* (Beverley Hills, Sage: 1983).

B. Latour, *Science in Action: How to Follow Scientists and Engineers Through Society* (Cambridge, MA: Harvard University Press, 1997).

J. Law and J. Hassard, *Actor Network Theory and After* (Oxford: Blackwell Publishers, 1999).

P. Macnaghten, M. Kearnes and B. Wynne, 'Nanotechnology, Governance, and Public Deliberation: What role for the Social Sciences?', *Science Communication*, Vol. 27, No. 2, December (2005), 1–24.

Nature, 'Going Public', *Nature*, 431 (2004) p. 883.

Prince of Wales, Comment, *Independent on Sunday* (11 July 2004), p. 25.

A. Rip, 'Technology Assessment as Part of the Co-Evolution of Nanotechnology and Society: the Thrust of the TA Program in NanoNed', paper presented at *Nanotechnology in Science, Economy and Society* (Marburg, 13–15 January 2005).

RNCOS (Research and Consultancy Outsourcing Services), *The World Nanotechnology Market*, report (1 August 2006): last accessed 20/12/06 from – http://www.rncos.com/Report/IM060.htm

A. Stirling, *Deliberate Futures: Precaution and Progress in Technology Choice* (London: Royal Institution of Chartered Surveyors, May 2005).

D. Taverne, 'Let's Be Sensible About Public Participation', correspondence, *Nature*, 432 (November 2004).

The Royal Society and The Royal Academy of Engineering, *Nanoscience and Nanotechnologies: Opportunities and Uncertainties*, RS Policy document 19/04 (July 2004).

The Royal Society, *Government Commits to Regulating Nanotechnologies But Will It Deliver?*, Press Release (25 February 2005). Available: http://www.royalsoc.ac.uk/news. asp?id=2976

UK Government, *The Government's Response to Better Regulation Task Force's report on Scientific Research: Innovation with Controls* (2003). Available: www.brtf.gov.uk/taskforce/responses%20new/Scienceresponse.doc

J. Wilsdon, B. Wynne and J. Stilgoe, *The Public Value of Science. Or How to Ensure that Science Really Matters* (London: Demos, 2005).

B. Wynne, 'Afterword', in Kearnes, M., Macnaghten, P. and Wilsdon, J. (eds) *Governing at the Nanoscale. People, Policies and Emerging Technologies* (London: Demos, 2006).

B. Wynne, 'Creating Public Alienation: Expert Cultures of Risk and Ethics on GMOs', *Science as Culture*, 10, 4 (2001) 445–481.

7
Public Acceptability of Hydrogen Fuel Cell Transport and Associated Refuelling Infrastructures

Tanya O'Garra, Peter Pearson and Susana Mourato

1. Introduction

There are many drivers for change in fuels and energy technologies today. The combustion of fossil fuels to generate energy has led to increasing local, regional and global air pollution, rising temperatures due to increasing carbon dioxide emissions in the atmosphere, and the emergence of energy security issues associated with declining resource availability and political instability in the Middle East, the main oil producing region in the world (IPCC, 2001). The search is on to find energy carriers that promise reduced emissions and stable supply. Such energy carriers, coupled with efficient technologies, may resolve many of the problems that the world faces as a result of the dependence on fossil fuels.

Hydrogen (hereafter, H2) as a fuel for transport is currently being tested in a number of demonstration projects worldwide. Small numbers of H2 buses, light-duty trucks and other fleet vehicles are to be found on streets across the globe, from Perth in Australia to Reykjavik in Iceland; from Tokyo in Japan to Vancouver in Canada. The aim of these demonstration projects is to gather real-world data that will help to assess the viability of H2-fuelled vehicles and supporting infrastructures, with a view to achieving eventual full commercialisation in the vehicle market.

As with the introduction of any new technology, however, several constraints require examination. A key area of concern expressed by some experts is that the public might reject hydrogen transport technologies or infrastructures because of associations with the 1937 Hindenburg airship disaster or the 'hydrogen' bomb (e.g. Foley, 2001; Adamson and Pearson, 2000). Given the history of public opposition and resistance to several new technologies, such as nuclear power, wind power, and genetically modified crops (Gamboa and Munda, 2006; Huffman *et al.*, 2007; Gaskell and Bauer, 2001; Surrey and Huggett, 1976), it is crucial to identify whether these concerns have any foundation.

Understanding public acceptability of H2 transport and associated H2 refuelling infrastructure will allow for potential objections to be identified early on, and addressed accordingly through public consultation and deliberation processes, information campaigns, further research and testing, product or safety modifications, etc. Here, 'the *public*' consists of anyone who might be affected by the large-scale introduction of H2 buses or the storage of hydrogen at refuelling stations in urban centres.

This chapter is a synthesis of results from three studies carried out between May 2004 and August 2006 which aimed to address the public acceptability of or resistance to H2 transport and associated refuelling infrastructure in large cities, particularly London. Specifically, the studies sought to identify existing levels of public awareness about H2 as a fuel for vehicles, attitudes and willingness to pay for the large-scale introduction of H2 buses, attitudes towards the storage of hydrogen at refuelling stations, and drivers for opposition to infrastructure development. These studies are described in detail in Section 3.

The findings reported here should provide a baseline picture of acceptability and preferences for hydrogen at the earliest stage of introduction of H2 vehicles and infrastructure. Future studies might refer to this baseline to understand how public acceptability and preferences for H2 transport technologies, in London particularly, have changed over time.

This chapter is structured as follows: Section 2 reviews the existing literature on acceptance of and preferences for hydrogen-fuelled vehicles. It also briefly reviews the literature on acceptance and preferences for other alternative-fuel vehicles, as well as the literature on opposition to local infrastructure developments. Section 3 describes each of the case studies presented in this chapter. Section 4 summarises findings from these studies, and Section 5 presents the discussion and conclusions.

2. Literature review

At present only a handful of studies have investigated public acceptability for H2 vehicles. These include the 'Public Acceptance of Hydrogen' study carried out in Iceland as part of the ECTOS (Ecological City Transport System) project, which assessed attitudes towards hydrogen amongst H2 bus users in Reykjavik. Results indicated that respondents were largely positive about the buses and that safety was not seen as an issue (Maack and Skulason, 2006; Maack *et al.*, 2004). Similar findings are reported in Haraldsson *et al.* (2006), who found that users of fuel cell buses in Stockholm were generally positive about the buses, and three quarters of them indicated that they 'felt safe' onboard the H2 bus. Altmann and Graesel (1998) also found that users of a H2 bus in Berlin were positive about hydrogen fuel, and not concerned about safety. Likewise, safety issues did not concern taxi drivers opting to use H2-fuelled taxis, in Mourato *et al.* (2004). Other small-scale studies (e.g. Lossen *et al.*, 2003; Dinse, 2000; 1999) further support these findings. Overall, these survey study results indicate that, at present, the public is not particularly troubled by hydrogen, and that attitudes towards H2 vehicles are largely positive.

Mourato *et al.* (2004) carried out the only reported study that specifically assessed *preferences* for hydrogen vehicles. They found that taxi drivers were willing to pay to drive H2 fuel cell (FC) taxis, and that willingness to pay (WTP) was determined by expectations of financial gain in the short-term, although in the long-term it was also influenced by environmental concerns. Apart from Mourato *et al.* (2004) no other studies appear to have estimated WTP for H2 vehicles.

Most of the existing literature on public preferences for new environmental transport technologies and fuels focuses on the demand for private alternative-fuel vehicles (especially electric vehicles).[1] In general, results from these studies suggest that environmental concerns are not important in the choice of transport technology. For example, Chiu and Tzeng (1999) found that concerns about emission levels were only a weak determinant of preferences for electric motorcycles. Similarly, Ewing and Sarigöllü (1998) found that the high potential demand for low emission private vehicles in Montreal was not determined by emissions, but rather by the price and performance of the vehicles.

As yet, no studies have reported economic values of the environmental benefits of H2 buses.[2] Although several studies estimate the environmental benefits of alternative-fuel buses (e.g. Karlstrom, 2005; Schimek, 2001), they tend to use indirect damage-cost approaches to estimate environmental benefits. To the best of our knowledge, the AcceptH2 study (see below) is the first study directly to estimate WTP for the environmental benefits of alternative-fuel buses.

Similarly, to the best of our knowledge, there are no studies of the public acceptability of H2 refuelling stations. However, there has been a considerable amount of empirical research over the past 30 years into the determinants of opposition to locally unwanted land uses such as nuclear waste sites, landfills, power plants and wind farms. This literature has been reviewed extensively in O'Garra (2005a). Opposition has been found to be more complex than had originally been assumed when opposing communities were characterised as shortsighted and self-interested, and opposition was usually explained simply as a NIMBYist ('Not-in-my-backyard') reaction (e.g. O'Hare, 1977). It is now more widely recognised that opposition is often associated with several interrelated factors including: the distance of the facility from residential areas, perceived risk, distrust of facility developers and distrust of facility siting procedures.

3. Overview of case studies

In this section we briefly describe the three survey studies that form the basis of this chapter. All three were coordinated by and/ or carried out at Imperial College's Centre for Energy Policy and Technology (ICEPT). More detailed results can be found in O'Garra *et al.* (2005), O'Garra (2005a), O'Garra (2005b).

3.1 AcceptH2

The AcceptH2 project was an EU-funded collaboration between five cities worldwide, where H2 bus trials were taking place: London (UK), Berlin (Germany), Luxemburg, Perth (Western Australia) and Oakland (US). Results for Oakland will not be presented here, as this study used a very different survey instrument and sampling approach; thus results are not considered strictly comparable.[3] The bus trials involved three H2 fuel cell buses running for two years in London, Luxembourg and Perth. The trial in Berlin involved one H2 bus running for 18 months.[4] The research consisted of a comparative study of public acceptability of H2 FC buses before *and* after their introduction, and an estimation of the economic value of their environmental benefits, carried out by the project partners.[5] To the best of our knowledge, the AcceptH2 study is the first study directly to estimate willingness to pay (WTP) for the environmental benefits of alternative-fuel buses (see below).

Survey design and data collection

Two sets of surveys were carried out for this study. The first set of surveys was administered to respondents in Berlin, London, Luxembourg and Perth *before* the trial of the H2 buses in each city (from henceforth it shall be referred to as the *ex ante* survey). It involved an extensive survey instrument, which was appropriately pre-tested using focus groups and pilot interviews, aimed at identifying knowledge and perceptions about hydrogen-fuelled transport, attitudes associated with the introduction of H2 vehicles in general, as well as H2 FC buses in particular, and preferences towards the introduction of H2 FC buses in each of the study-cities. Socio-economic characteristics (e.g. age, income) and general environmental attitudes, knowledge and behaviour of respondents were also identified.[6]

The *ex post* questionnaire, carried out in Berlin, London, Luxembourg and Perth after the H2 buses had been running for about six months, was adapted from the longer *ex ante* questionnaire and streamlined to make it shorter. In addition, Berlin and Luxembourg carried out small samples of interviews with respondents on-board the H2 buses.

The *ex ante* surveys were conducted with residents of Berlin, London, Luxembourg and Perth between July 2003 and February 2004. The Berlin and Luxembourg surveys were conducted only with bus users, whilst the London and Perth samples included non-bus users too. The *ex post* surveys were carried out approximately one year later between July 2004 and February 2005. Table 7.1 summarises the data collection for each city.[7]

Sample characteristics

This study was aimed primarily at standard bus users, although as noted in the table, the London and Perth surveys also captured a significant proportion of non-bus users. However, for reasons of space, only bus user results will be presented in this chapter. Where consid-

Table 7.1   Summary of survey data collection for the AcceptH2 project

	Berlin	**London**	**Luxembourg**	**Perth**
Ex ante survey				
Total sample size	344	414	300	300
No. bus users	344	306	300	147
% bus users	100	74	100	49
Ex post survey				
Total sample size	263	300	301	300
No. bus users	263	249	301	165
% bus users	100	83	100	55

Table 7.2   Socio-demographic characteristics and attitudes of AcceptH2 study
bus user samples

Variables	Berlin sample (n=537)	London sample (n=714)	Luxembourg sample (n=554)	Perth sample (n=600)
Gross annual household income (national currency)	31,788	44,643	43,202	51,924
Gross annual household income (Euros)[a]	31,788	64,700	43,202	31,092
Sex (% male)	44.8	43.0	35.2	42.0
Age (mean)	50.2	39.5	47.3	43.8
University education (% respondents)	25.6	48.2	33.8	40.1
Employed (% respondents)	47.1	80.1	56.0	56.7
Car ownership (% households owning car)	56.8	69.2	79.9	90.7
Fumes from existing buses considered bad or very bad (% respondents)	17.4	50.1	16.1	22.6
Environmental attitude (range 1–5)[b]	3.98	3.91	4.08	4.01

[a] Mean attitude to statement: '*Solving environmental problems should be one of the top three priorities for public spending in London*' (from 1-strongly disagree to 5-strongly agree)
[b] Concerted using April 2005 exchange rate (*Source:* www.xe.com/ucc)

ered relevant, non-bus user statistics for London and Perth will also be commented on for comparative purposes. Results and analysis of whole-sample data for London and Perth can be found in O'Garra *et al.* (2007).

Bus user sample characteristics and attitudes for AcceptH2 surveys are presented in Table 7.2. A bus user is defined as *someone who has used a bus at least once in the past 12 months*. Overall, results indicate that the samples from each city have very different socio-economic characteristics. For example, Berlin has a significantly higher proportion of males in the sample compared to the other cities. Education levels are also very different between samples. The significant discrepancy between income levels is, however, perhaps most notable. As the figures show, average income levels (in Euros) for the London sample are about double the Berlin and Perth samples' average income, and

€20,000 above the next highest income (in Luxembourg). Due to a lack of sufficient data on the characteristics of bus user populations in each city, it was not possible to weight the data accordingly.[8] Therefore generalisations of the un-weighted sample results to the overall city bus user populations should be interpreted with caution.

3.2 Acceptability of hydrogen refuelling stations in London

This study formed part of the UK Engineering and Physical Sciences Research Council (EPSRC)-funded project, *The Development of a Hydrogen Energy Infrastructure for London.*[9] This project involved the development and estimation of a model of H2 infrastructure development for refuelling fleet vehicles in London, incorporating technical, economic and social factors. This chapter presents results from the acceptability surveys that were carried out in London, as part of this wider project.

Survey design and data collection

This study involved one questionnaire, which aimed to establish: respondents' past and expected future length of residence in the area, attitudes towards their local area, including the petrol station, and past participation in local planning issues. It also established existing knowledge about H2 as a fuel for vehicles, attitudes towards H2 vehicles, and attitudes to H2 storage at existing refuelling stations. Note that attitudes were explored before and after giving respondents information about hydrogen. The questionnaire also elicited information on respondent socio-economic characteristics and environmental attitudes, knowledge and behaviour.

The sampled population included households near existing commercial refuelling stations in London, as it was expected that local communities would bear the main external cost of hosting a H2 refuelling facility. Because this study was targeted at households, refuelling stations with very low residential population density were dropped (for more detailed information on sampling procedure see O'Garra (2005a) and note).[10]

Almost 800 questionnaires were distributed in five London boroughs[11] between 18th September and 18th December 2003. One third of these surveys (31 per cent; n=245) were picked up the following week, and a further sixth of the sample (16 per cent; n=131) were returned by post. The overall response rate is a fairly high 47 per cent. Of the 376 returned surveys, less than one tenth (30) were incomplete or not useable, and were dropped from the analysis.

Table 7.3 Socio-demographic characteristics of samples from acceptability of H2 infrastructure studies

Variables	H2 Infrastructure Survey (n=346)	BP-IC Infrastructure Survey – Havering (all survey phases) (n=1014)	BP-IC Infrastructure Survey – Bromley (all survey phases) (n=1010)
Gross annual household income (mean in £)	40,875	30,833	35,724
Sex (% male)	51.3	48.2	49.1
Age (mean) (a)	44.5	51.8	51.5
University education (% respondents)	50.5	20.4	34.0
Employed (% respondents)	68.5	53.7	56.3
Car ownership (% households owning car)	80.2	83.7	84.0
Respondent's home is <200m from nearest refuelling station (% respondents)	88.6	16.2	19.0
Respiratory illness in respondent's household (% respondents)	–	32.0	29.7
Environmental attitude (range 1–5)[a]	3.69	4.06	4.00
Interest in new technologies (range 1–5)[b]	–	2.85	2.97
Mean level of concern towards storage of H2 at local refuelling station (from 1-not at all concerned to 5-very concerned)	2.98	2.77	2.80

[a] Mean attitude to statement: 'Solving environmental problems should be one of the top three priorities for public spending in London' (from 1-strongly disagree to 5-strongly agree)

[b] Mean agreement with statement: 'I am always reading about the development of new technologies' (from 1-strongly disagree to 5-strongly agree)

Sample characteristics

Key socio-economic and attitudinal characteristics of the H2 Infra-structure survey sample are presented in the first column of Table 7.3. Mean annual household income before tax is marginally higher than the London average income of £38,376 per annum,[12] but the difference is not statistically significant. University education levels on the other hand are significantly higher (at 1 per cent level) than the London average, with almost half the sample having completed university degrees compared to the London average of 25 per cent (ONS, 2004). Notably, both income and education levels are markedly higher amongst this survey sample compared to the BP-IC survey samples shown in the next two columns of Table 7.3 (discussed in more detail in the following section). This indicates possible self-selection of respondents to the questionnaire, and hence generalisations of the un-weighted sample results to the overall London population should be interpreted with caution.

3.3 Longitudinal analysis of attitudes in the London borough of Havering

This is an ongoing survey-based study funded by British Petroleum, and coordinated by the Imperial College Centre for Energy Policy and Technology. This study was motivated partly by the rejection of plan-ning permission for the Hornchurch hydrogen refuelling facility, in the borough of Havering in London. The planning application was rejected in June 2003 because of opposition from local neighbours and council-lors. One year later, the process went to public enquiry and the station was approved on the 28th July 2004. It is now fully operational.

This case highlighted the value of investigating the impact of the Hornchurch station development process on public knowledge and attitudes towards hydrogen in the areas surrounding the Hornchurch facility. The borough of Havering was taken as the area of interest, and surveys carried out there periodically every six months in order to chart attitude changes over time. In addition, a control borough was included in the study (Bromley), in order to assess whether changes in Havering could be attributed to the Hornchurch refuelling station.[13]

Survey design and data collection

Design of this survey drew from the authors' experience associated with the AcceptH2 and H2 Infrastructure studies; thus, a number of questions are common to all surveys. This facilitates comparison of results. Following adequate pre-testing of the survey, the final ques-

Table 7.4 Summary data collection statistics

Survey phase	Survey period	Sample sizes		
		Total	Bromley	Havering
Phase 1	May/June 2004	411	198	213
Phase 2	Nov/Dec 2004	399	205	194
Phase 3	June/July 2005	402	201	201
Phase 4	Dec/Jan 2005	407	204	203
Phase 5	June/July 2006	405	202	203
Total		2024	1010	1014

tionnaire elicited information on: existing knowledge about H2 as a fuel for vehicles, attitudes towards the H2 vehicles, and attitudes to H2 storage at existing refuelling stations. It also elicited information on environmental attitudes, knowledge and behaviour, as well as knowledge about hydrogen fuel and other alternative fuels. No information about hydrogen was provided in this survey.

Survey respondents were contacted by a market research company (Carrick James Research) using random-telephone dialling, based on telephone directories for the boroughs of Havering and Bromley. Table 7.4 summarises the data collection statistics for each survey phase.

Sample characteristics

Sample characteristics for the boroughs of Havering and Bromley are presented in the last two columns of Table 7.3. These descriptive statistics have been obtained by aggregating data from all the survey phases (from summer 2004 (Phase 1) to summer 2006 (Phase 5)). It must be noted that sample characteristics are quite consistent across survey phases.

As Table 7.3 shows, the Havering and Bromley samples differ quite significantly from each other in terms of income and education levels, with Bromley respondents being generally more educated and more affluent. These figures closely reflect population statistics for the two boroughs. However, the survey samples are more highly educated and have a higher average income than the populations they were sampled from. This has been a recurring trend throughout all the studies reviewed in this chapter, and most likely indicates a survey selection bias, whereby individuals with these characteristics are more willing to complete a survey of this kind. Using population statistics for each borough, the samples have been weighted to control for age, education and car ownership levels (using population statistics from Census 2001) and population income levels (using LATS 2001).

4. Summary of findings

This section presents selected findings from the three studies of acceptability and preferences for hydrogen transport and refuelling infrastructure, under key headings.

4.1 Public awareness of hydrogen

How aware were the public about ongoing developments relating to hydrogen? Before being given information about hydrogen and fuel cells, respondents in all surveys were asked whether they knew that car companies were developing H2 vehicles.

Results from the *ex ante* AcceptH2 surveys (carried out between May 2003 and Feb 2004) indicate that Berlin bus users were by far the most informed – three quarters (72 per cent) had heard about the development of H2 vehicles; bus users in Perth, Luxembourg and London were much less well informed – only around half had heard about H2 vehicles (54 per cent, 51 per cent and 48 per cent respectively). This difference is not unexpected given that Berlin has had prior experience of H2-fuelled vehicle trials. Berlin was host to one of the world's first H2 fleet vehicle demonstration projects, which took place in the 1980s. This consisted of a four-year trial of ten H2-powered medical vehicles. Further H2-related demonstration projects followed in the late 1990s (e.g. the BMW Clean Energy World Tour 2001), and there was a H2 bus trial in the year 2000. This experience with H2 vehicles probably explains the higher awareness levels of Berlin respondents.

Six months into the trial of H2 buses in each of these four cities (*ex post* AcceptH2 surveys), awareness about H2 vehicles had increased in Luxembourg and Perth only, by about 16 per cent in both cases (see O'Garra, 2005b for full results). Additionally, Perth bus users were the most aware of the H2 bus trials taking place in their city (three fifths (59 per cent) said they knew about the trial), followed by Luxembourg (one half (51 per cent) and Berlin (45 per cent)). London bus users were the least likely to have heard about the bus trials, with only a fifth (20 per cent) of bus users having heard of them.

Although it is evident that bus trials of a small number of H2 buses (e.g. three in London) will be more visible in smaller cities such as

Luxembourg (population 450,000) or Perth (population 1.5m), compared to larger cities such as London (population just under 8m) or Berlin (population 4.5m), it is suggested that the levels of awareness in each city may also reflect the extent of public outreach and the information campaigns that accompanied these demonstration projects. In Perth, for example, there were extensive efforts which included: brochures on board buses, cut-out cardboard model buses sent to schools, a dedicated website, a TV programme, articles in the local newspapers, radio interviews, and conferences associated with the launch of the H2 bus trials. In Luxembourg, the local public transport operator made radio commercials and the fuel cell buses participated in the Luxembourg Spring Fair. Such efforts were not seen in London or Berlin. This may also help to explain the lack of increase in awareness levels.

The BP-IC study further expands upon the findings in the AcceptH2 study, by charting changes in awareness about H2 vehicles in several London boroughs, between summer 2004 and summer 2006. Results from this study indicate that awareness about H2 vehicles has increased only marginally in the boroughs of Bromley and Havering. More to the point, awareness levels appear to be fluctuating in both boroughs over time. For example, results for Havering indicate that there was a large increase in the number of residents who had heard about H2 vehicles (from 37 per cent to 53 per cent) between summer 2004 and summer 2005. This may have been associated with the construction of the refuelling station and associated media coverage, and/or word-of-mouth communication between Havering residents. After this initial increase, awareness levels stabilised for a period. However, from the results for summer 2006 – one year after the construction of the H2 refuelling station – it appears that more than one sixth of the sample (15 per cent) had 'forgotten' about hydrogen.

In cognitive psychology it is generally accepted that individuals experience accelerating 'forgetting curves' with respect to new messages or information (see Eagly and Chaiken (1993) for a review of relevant studies). Thus one study showed that as little as two weeks after exposing individuals to new information, 40 per cent were unable to recall that topic (Watts and McGuire, 1964: 237). Only repeated exposures ensure an almost 100 per cent recall rate (Eagly and Chaiken, 1993). On the basis of this previous research, it is tentatively suggested that the Havering and Bromley residents have not been exposed to sufficient information about hydrogen in order to ensure high enough recall rates, hence the varying awareness levels over time.

Analysis of the determinants of awareness[14] indicates that, in all cities, men are much more likely to have heard about H2 vehicles than women. This is perhaps unsurprising since men tend to have greater interest in cars and new technologies. Other determinants of knowledge include: university education, interest in new technologies, income and environmental attitude.

4.2 Public attitudes to H2 vehicles

Existing attitudes towards a scenario of large-scale introduction of H2 vehicles in the respondents' cities, were elicited in the *ex ante* AcceptH2 surveys.[15] The highest support levels were found amongst Berlin bus users (almost seven tenths (69 per cent) support H2 vehicles), and the lowest levels of support were found amongst London respondents (two fifths – 39 per cent) of bus users, and only one quarter (25 per cent) of non-bus users support the introduction of H2 vehicles). In keeping with findings about awareness about H2 as a fuel for vehicles, presented in the previous section, most London respondents simply indicated that they 'need more information' (more than half – 57 per cent) of the bus users, and nearly seven tenths (69 per cent) of the non-bus users). Similar results are reported in the H2 Infrastructure study where one third (33 per cent) of respondents were found to support this scenario, whilst a three fifths majority (63 per cent) indicated a need for more information. Opposition was negligible in both cases.

Results presented in O'Garra *et al.* (2005) indicate that – for London respondents – support was largely determined by prior knowledge of H2 vehicles. This is not surprising, as people often require some awareness or knowledge of an issue, before giving an opinion on it. The fact that respondents tend to *support* H2 vehicles on the basis of existing knowledge probably reflects the positive nature of the available information. O'Garra *et al.* (2005) suggest that if the publicly available information were mostly negative, support levels would probably be lower.

These findings are further supported by the results of the BP-IC study, which also asked respondents for their attitude towards the large-scale introduction of H2 vehicles in London. Multinomial regressions were used to identify which variables influenced respondents' attitudes, and also whether there had been a statistically significant change in attitudes over time in Havering. The base category used in the regression is '*need information*'. Hence, all results are interpreted as comparisons with this alternative. Table 7.5 presents the results of this regression.

These results confirm previous findings that prior knowledge about hydrogen as well as direct experience of a H2 bus have a very major

139

Table 7.5 Multinomial regression on attitude to large-scale introduction of H2 vehicles: pooled model results for Havering and Bromley

	SUPPORT		OPPOSE		INDIFFERENT	
	Coefficient	t-ratio	Coefficient	t-ratio	Coefficient	t-ratio
Income (divided by 10,000)	0.06**	2.32	−0.16	−1.49	−0.07*	−1.68
Male (1=yes)	0.72***	6.12	0.38	1.19	0.77***	4.89
Age (in years)	0.01**	2.28	0.02*	1.70	−0.00	−1.07
University education (1=yes)	0.21	1.44	−0.72	−1.01	−0.69**	−2.45
Owns a car (1=yes)	−0.47***	−3.55	−0.58*	−1.78	−0.93***	−5.62
Environmental attitude (range 1–5)[a]	−0.03	−0.57	−0.24*	−1.70	−0.27***	−3.75
Interest in new technologies (1–5)[b]	0.22***	4.64	−0.28**	−1.97	−0.14**	−2.12
Has heard about H2 vehicles (1=yes)	1.45***	12.09	−0.45	−1.23	0.08	0.50
Has used a H2 bus (1=yes)	1.60***	2.55	−36.79	−0.00	0.61	0.58
Survey phase indicator (1=summer '04 to 5=summer '06)	0.02	0.32	−0.20	−1.35	0.18	0.27
Lives in Havering (1=yes)	0.19	0.76	−1.11*	−1.69	0.10**	2.53
Havering resident*survey phase (interaction variable)	−0.02	−0.21	0.43**	2.13	0.03	' 0.31
Constant	−2.80***	−7.17	−1.32	−1.26	−0.19	−0.36
LR Chi2 (36)	569.54***					
Log-likelihood	−1899.17					
Pseudo-R2	0.13					
No. observations	2152					

* Significant at the 10% level; ** at the 5% level; *** at the 1% level

[a] Mean attitude to statement: '*Solving environmental problems should be one of the top three priorities for public spending in London*' (from 1-strongly disagree to 5-strongly agree)

[b] Mean agreement with statement: '*I am always reading about the development of new technologies*' (from 1-strongly disagree to 5-strongly agree)

positive influence on support for H2 vehicles overall. In fact, the marginal impact of these variables on support levels[16] is about 30 per cent for both explanatory variables: for example, direct experience of a H2 bus increases support by almost one third. Interestingly, prior knowledge of hydrogen does not appear to be a requisite for respondents to oppose H2 vehicles: opposition may not be related to specific perceptions of hydrogen *per se*, but to other issues and concerns.

Also worth noting is the increase in opposition towards H2 vehicles (significant at the 5 per cent level) in Havering compared to Bromley, independent of all other factors, suggested by the positive sign on the interaction variable (Havering resident*survey phase).[17] In contrast, support levels have not changed over time.

All of these results confirm previous findings (O'Garra *et al.*, 2005; O'Garra, 2005a). In summary: 1) knowledge and experience of hydrogen technology are key drivers for support, and 2) opposition is *not* based on knowledge or awareness about hydrogen technologies. These findings should serve to emphasise the value of providing members of the public with direct experience of new technologies such as hydrogen, and the value of continued information-provision, to ensure adequate recall and hence, better informed decision-making.

4.3 Public attitudes to hydrogen buses

After being given information about H2 fuel cells and the H2 bus trials, interviewees in the *ex ante* AcceptH2 surveys were asked if they thought the H2 bus demonstration projects were a good idea. The rate of unconditional support for the trials was overwhelming (90 per cent overall) and opposition insignificant (1 per cent). However, respondents were much more cautious about the *large-scale* introduction of H2 buses in their cities: Figure 7.1 shows the distribution of these results.

Results show that while almost half (46 per cent) of bus users overall unconditionally support the large-scale introduction of H2 buses in their cities, almost the same amount (44 per cent) indicated that their support was conditional on the results of the trials and the resolution of safety issues. Opposition – mostly from safety concerns – was negligible (3 per cent). Bus users in Berlin and Luxembourg showed the highest levels of unconditional support, whereas Perth bus users expressed the highest levels of conditional support. Overall, these results confirm findings from previous studies that public attitudes towards H2 vehicles are largely positive, and safety concerns are not thus far an issue.

Figure 7.1 Attitudes towards the large-scale introduction of hydrogen buses

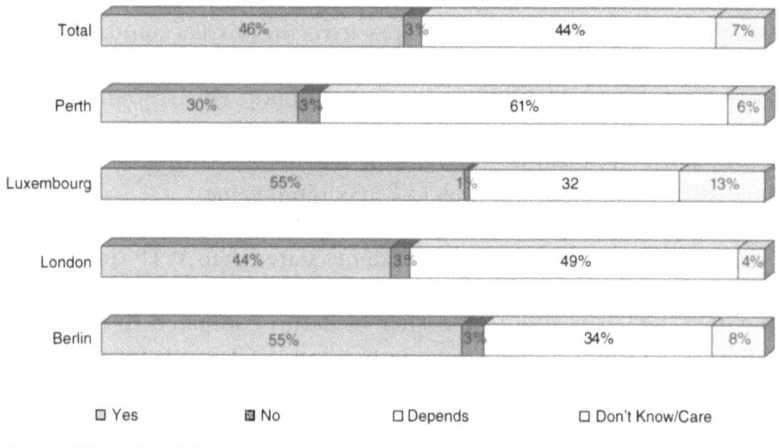

☐ Yes   ⊞ No   ☐ Depends   ☐ Don't Know/Care

Source: O'Garra (2005b)

Six months into the bus trials, results of *ex post* AcceptH2 surveys indicate that unconditional support for the large-scale introduction of H2 buses has increased in all cities, by about one fifth overall (i.e. 20 per cent more respondents support H2 buses).

Influences on unconditional support include prior knowledge about H2 vehicles, which is a very strong driver for support in Berlin, Luxembourg and Perth. As noted in the previous section, this positive relationship between knowledge and support is likely to be a direct consequence of the largely positive nature of the available information about hydrogen in the public domain (O'Garra *et al.*, 2005).

Interestingly, respondents who had used a H2 bus were not more likely to offer unconditional support for their large-scale introduction. This may appear to contrast with findings presented in Section 4.2, where support for H2 vehicles was found to be very strongly (positively) influenced by direct experience of a H2 bus (BP-IC study). In fact, results from the AcceptH2 study indicate that respondents who have used a H2 bus are more likely to express support, *conditional* on results of the trials.

4.4 Willingness to pay to support hydrogen buses

Were bus users actually willing to pay extra for hydrogen buses? Respondents to the AcceptH2 study were presented with the following scenario: '*Suppose there was a proposal to substitute the buses in the [city] transport system for hydrogen fuel cell buses.*' They were then asked whether they would be willing to pay extra on top of a standard single €2.00 ticket in Berlin, £1.00 ticket in London, €1.20 ticket in Luxembourg and AU$2.00 in Perth, to support this scenario. Those who were willing to pay (WTP) were asked to indicate how much, using a payment ladder (a series of payments starting at zero and increasing by discrete amounts to a maximum value).

Overall, most respondents indicated a positive WTP, except in Berlin where just over half of the respondents stated zero WTP (in both *ex ante* and *ex post* surveys). This is probably due to the price of a single bus fare, which is about €1 higher in Berlin compared to the other cities in the study. Table 7.6 presents mean WTP results for each city. All values have been converted to Euros (€), as valued in April 2005. In addition, in order to make values comparable, they have been adjusted to the cost of living in each city, with Berlin as the baseline (so all values are in terms of Berlin Euros).[18]

As results show, mean WTP extra bus fare is highest in Luxembourg, by an average of €0.10. Otherwise, WTP is relatively similar across cities, indicating that preferences for H2 buses are very similar across the world. Regression analyses carried out on WTP data indicate that, despite these similarities, the drivers for WTP differ quite significantly across cities – however, what all cities have in common is that WTP did not increase significantly six months after the H2 bus trials. This might seem a surprising result, given that unconditional support for H2 buses was found to increase significantly (noted in previous section). It is likely that respondents consider their responses more carefully when asked for potential payments, and hence it is considered that this result reflects more appropriately respondents' feelings about H2 buses in their respective cities.

Table 7.6   Mean willingness to pay extra bus fare for introduction of H2 buses

	ex ante	*ex post*
Berlin (€)	0.30	0.33
London (€)	0.28	0.30
Luxembourg (€)	0.43	0.43
Perth (€)	0.35	0.30

Note: all values have been adjusted to the cost of living

4.5 Attitudes to hydrogen refuelling stations by individuals living close by

What do respondents feel about H2 refuelling stations? Existing attitudes towards the stations (location not specified) were elicited from respondents to the H2 Infrastructure study, by asking them: *'What are your initial feelings about the storage of hydrogen at existing refuelling stations in London?'* using a simple *'support/oppose/need more information/don't care'* scale. A significant three quarters (74 per cent) of respondents indicated the need for more information, whilst one tenth (13 per cent) expressed support for this development. Thirteen respondents (3.7 per cent of sample) opposed the storage of H2 at refuelling stations in London: all gave safety reasons for their attitudes. Figure 7.2 shows the distribution of responses, and compares this to the distribution of attitudes towards a scenario of large-scale introduction of H2 vehicles in London.[19]

As Figure 7.2 shows, respondent attitudes are more positive towards the introduction of H2 vehicles in London, than about the storage of hydrogen in refuelling stations in London. A larger proportion of respondents also says that they would need more information about H2 storage than about H2 vehicles.

The issue of H2 storage at refuelling stations in London was then brought closer to home, by asking: *'Suppose a proposed hydrogen refuelling facility development were to take place* in three months time *at your local petrol station. Would you in principle support it, oppose it, need more*

Figure 7.2 Comparing existing attitudes to H2 vehicles and H2 storage at refuelling stations in London

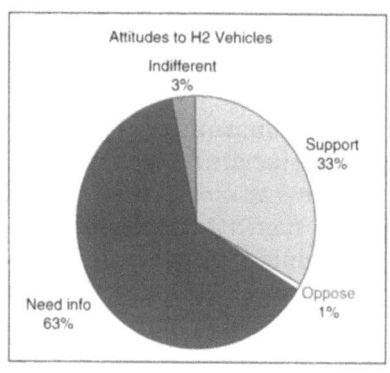
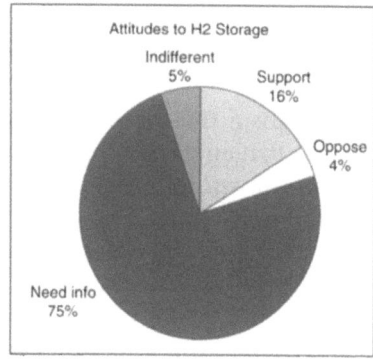

Source: O'Garra (2005a)

Figure 7.3   Distribution of attitudes towards H2 storage development going ahead in three months' time at the respondent's local refuelling station

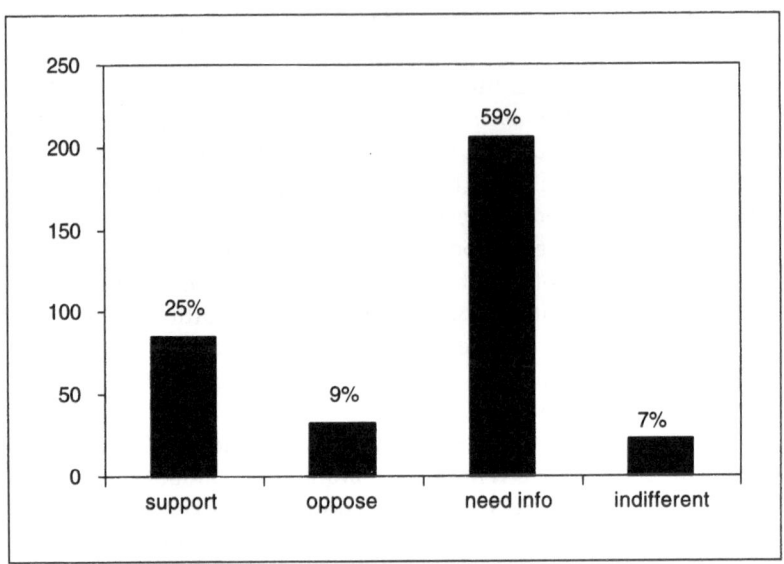

Source: O'Garra (2005a)

information to make a decision or are you indifferent/don't care?' Figure 7.3 presents the distribution of results for this question.

Clearly, the majority of respondents (three fifths) indicated that they *'need more information'* and support levels were moderate, with a quarter of respondents supporting such a development. One tenth of respondents (n=33) said they would oppose it. This is significantly higher than the small proportion (only 4 per cent) who indicated that they opposed H2 storage at non-specified refuelling stations in London. These 'opposers' are the individuals that are particularly interesting.[20]

Analysis of the determinants of the determinants of opposition[21] reveals that opposition to H2 storage at a respondent's local refuelling station was strongly explained by concerns about H2 safety, lack of trust in safety regulations and past participation in local planning decision-making efforts. In addition, older, university-educated respondents with less interest in environment issues were more likely to oppose. Also, as expected, the closer the respondent lives to the proposed H2 refuelling station, the more likely they were to oppose the development.

It is particularly interesting to note that respondents agreeing with the statement *'I'd prefer it if hydrogen were stored at a refuelling station elsewhere in London'* were more likely to oppose local H2 storage (significant at the 1 per cent level). This finding tentatively supports the traditional – and unpopular (e.g. Wolsink, 2000; Lober and Green, 1994) – NIMBY explanation of self-interested opposition to facility siting. Perhaps surprisingly, years of residence in the area and expected future years of residence have no effect on likelihood of opposition.

Finally, it appears that prior awareness about hydrogen was not required for a respondent to oppose local storage of the fuel. This confirms findings presented in previous sections, and further confirms that those in opposition may not be opposed to hydrogen and may have other related concerns and issues.

4.6 Public attitudes to hydrogen refuelling stations over time

How do attitudes change as time passes? As noted in Section 3.3, the BP-IC study aimed to assess changes in attitudes over time, in three London boroughs. The borough of interest is Havering, the location of the Hornchurch H2 refuelling station; the borough of Bromley is used as the control, given that it has had no direct experience of H2 vehicle trials or H2 infrastructure development. Thus changes in attitudes in Havering are compared to changes in Bromley, in order to understand whether the Hornchurch refuelling station development has had any particular impact on awareness or attitudes.[22]

Results indicate that there have been no significant changes in support or opposition to local H2 storage over time in Havering or Bromley. Only indifference appears to have increased overall in both boroughs (although this is only significant at the 10 per cent level). Individual regressions for each of the three boroughs in this study confirm that indifference is increasing – nothing else appears to be changing.

Drivers for opposition were not clearly identified through this study. This could be either because individuals who oppose local H2 refuelling station developments are very different from each other, and hence no common characteristics can be identified, or because the survey failed to identify their key traits (e.g. personality types, political affiliations, etc). This is an area that would benefit from in-depth qualitative examination.

4.7 A closer look at opposition ...

Knowledge levels

Our regression results suggest that in most cases, the likelihood of opposition – whether in relation to the large-scale introduction of H2 vehicles or to local H2 storage – is not related to prior awareness of H2 vehicles. As noted, this suggests that opposition is not related to specific perceptions about H2, but to other issues and concerns.

Willingness to undertake action

Almost all respondents in the London H2 Infrastructure study indicated that they would be willing to spend time on opposition activities to prevent the development going ahead. In particular, four people were identified as potential organisers of opposition efforts (they indicated that they were willing to organise meetings and solicit signatures, whilst the rest were willing simply to sign petitions and turn up at meetings). The potential organisers are of particular interest as they mobilise opposition and create the opportunities for other opposers to voice their concerns (e.g. in local meetings or hearings with H2 developers). Although generalisations cannot be based on four respondents, it may be relevant that all four had university degrees and prior knowledge about H2 vehicles. Furthermore, all four *'strongly agree'* with the statement *'I'd prefer it if hydrogen were stored at a refuelling station elsewhere in London'*, and were all 100 per cent certain about their attitudes to H2 refuelling stations, so the potential organisers of opposition in this sample were knowledgeable, educated individuals with strong and certain attitudes towards the local development of H2 facilities.

In contrast to their willingness to actively engage in opposition efforts, only half of the opposing respondents indicated that they would be willing to make a donation towards a local opposition group, to support their efforts. This suggests that, on the whole, respondents may prefer to actually engage directly in opposition activities than indirectly, by donating money to opposition groups. There are several reasons why this might be the case: firstly, it may be that individuals prefer to make time rather than monetary trade-offs, if money is more highly-valued than time. Secondly, respondents might be behaving *strategically* whilst completing the questionnaire, by refusing to state monetary values associated with their opposition to the H2 storage development. Thirdly, it is possible that opposing individuals get private satisfaction out of engaging in opposition efforts. Participation may, for example, contribute to a sense of identity, self-worth, control, etc. This is an area worth further investigation.

Past experiences in opposing planning permission

Results from the London H2 Infrastructure study indicate that respondents who opposed local H2 storage developments tended to have past experience in soliciting signatures to petitions about a local planning issue. This suggests that opposers in the sample tended to be active in local planning efforts.

Weak determinants of opposition

Finally, it is worth noting that in all studies, the determinants of opposition are weak (the coefficients are very low) and not highly significant (most are significant only at the 10 per cent level and none at the 1 per cent level). This could be either because the characteristics of opposing individuals are very different, or because the survey failed to identify their key common traits. This is an area that would benefit from in-depth qualitative examination. It is striking that individuals who oppose local H2 developments do not necessarily know anything about hydrogen, or other new technologies for that matter. Thus, it appears that there are likely to be other reasons lying behind their attitudes.

5. Conclusions

This chapter presents results from three major survey studies carried out between May 2004 and August 2006, which aimed to address the public acceptability of or resistance to H2 transport and associated refuelling infrastructure in large cities, particularly London. In particular, these studies aimed to establish baseline social perceptions, knowledge and attitudes, in order that future public experiences with hydrogen might be interpreted in the light of these findings.

Overall, results from all studies indicate that public awareness about hydrogen as a fuel for vehicles was quite low in most cities surveyed, with approximately half of respondents in London, Luxembourg and Perth having ever heard of a H2 vehicle. Berlin was the exception: three quarters of respondents to the AcceptH2 survey in Berlin had heard of H2 vehicles. This is explained by Berlin's prior experience with H2 vehicle demonstration projects throughout the 1980's and up until 2000, during which year there was a H2 bus trial in Berlin.

Regression analysis indicates that six months into the bus trials, awareness levels in Berlin and London had not changed significantly, whereas there had been a statistically significant increase in public awareness of H2 vehicles in Luxembourg and Perth. These results suggest that the H2 bus trials have had a higher impact on public awareness in Luxembourg and Perth, whilst in Berlin and London the bus trials have been less successful in this respect (however, Berlin bus users were initially the most informed, so the potential for increase was not as significant as for London). Notably, Perth and Luxembourg had extensive information campaigns associated with the trials, whilst Berlin and London did not.

However, results from the BP-IC longitudinal study suggest that – despite the construction and visibility of the Hornchurch refuelling facility in Havering – awareness about hydrogen as a fuel for vehicles has only increased marginally in the past two years, and moreover, knowledge levels have fluctuated quite significantly over time. It is possible that the lack of continued media coverage about the Hornchurch facility (now built and functioning) has caused people to 'forget' about hydrogen. It would be interesting at this stage to investigate awareness levels in Perth and Luxembourg, now that the media coverage associated with the bus trials has largely ceased. It is possible that recall rates are similarly seen to decline, as with the BP-IC respondents.

Attitudes towards H2 vehicles are largely positive in all cities (opposition is negligible), with support levels highest in Berlin and lowest in London. This finding closely parallels knowledge levels in these cities (i.e. London respondents were the least aware about H2 vehicles, whilst Berlin respondents were the most aware). With this in mind, it is perhaps unsurprising to find that regression analysis reveals prior awareness to be the main driver for support for both the large-scale introduction of H2 vehicles and of H2 buses.

Six months after the H2 bus trials had begun unconditional support for the large-scale introduction of H2 buses had increased significantly in every city. However, increased unconditional support is not matched by increased willingness to pay for the large-scale introduction of H2 buses in most cases. Estimated WTP extra bus fare from the *ex ante* survey is not significantly different to that from the *ex post* survey (average of €0.35 for bus users in all cities). Thus it appears that, although attitudes towards H2 buses have become more positive in all cases, economic preferences (estimated as trade-offs between money and environmental improvements) have not changed much overall. This is hardly surprising: while people may express positive attitudes towards many good causes, in practice, they will only be willing to financially support a few of those.

Attitudes towards the storage of hydrogen at existing refuelling stations in urban centres, was explored only in London. Results show that attitudes are less positive towards H2 storage facilities, than towards H2 transport (discussed above). More respondents indicate that they 'need information' in order to give an opinion and opposition is no longer negligible. In particular, it is interesting to note that knowledge about hydrogen is not related to opposition in any way. Thus, those who oppose H2 facilities do not necessarily know anything about hydrogen. This suggests that they may be concerned about other issues. It is also worth mentioning that there are no strong explanatory factors for opposition likelihood. Coefficients on variables influencing opposition likelihood are low, and those that are significant are mostly so at the 10 per cent level. This raises the question: who are these opposing individuals? Surely they must have some characteristics in common? This would require further investigation, probably through qualitative examination of opposing individuals.

Overall, the findings reported in this chapter are largely positive. The public is not particularly concerned about hydrogen, and safety concerns rarely emerge as an issue in any of the studies. What is more prominent is the lack of knowledge about hydrogen. Given the evidence presented here – that knowledge seems to be a strong driver for support for the technology – the lack of public information might be considered questionable from a policy perspective. The question is: will a negative news flash cause the public to run screaming from hydrogen technologies?

Notes

1. A comprehensive review of studies of preferences for alternative fuel vehicles can be found in O'Garra (2005a).
2. Except for O'Garra and Mourato (2006) which analyses WTP for H2 buses using quantile regression analysis, using the London AcceptH2 data (see below) only.
3. Information on the Oakland study can be found on www.accepth2.com
4. Details about these H2 bus demonstration projects can be found on Fuel Cell Bus Club website (http://www.fuel-cell-bus-club.com/index.html), Fuel Cell Today website (http://www.fuelcelltoday.org) and International Energy Agency website (http://www.eere.energy.gov/hydrogenandfuelcells/hydrogen/iea/case_studies.html).
5. We acknowledge all the AcceptH2 project partners: David Hart (Imperial College London), Matthias Altmann and Patrick Schmidt (Ludwig-Bölkow-Systemtechnik GmbH, Germany), Lisa Garrity (Murdoch University,

Western Australia), Simon Whitehouse (Dept of Planning and Infra-structure, Western Australia), Cornelia Grasel, Anne Stevcevski and Judith Zell (University of Wuppertal, Germany). All partners contributed towards the development of the final survey, and data collection and analysis was carried out by respective partners. For more information on the acceptH2 project, see the project website: www.accepth2.com

6. All of the questionnaires (*ex ante* and *ex post*) established: bus usage, atti-tudes towards existing buses in each city (except in the Perth *ex post* survey, and the on-board survey), knowledge about H2 FC vehicles (except in the *ex post* on-board survey), and attitudes towards the potential large-scale introduction of H2 buses in each city. Respondents were then presented with neutral and balanced information on the advantages and disadvan-tages of H2 as a fuel for transport, and a brief description of the H2 bus trials taking place in each city. After giving respondents this information, attitudes towards the trial and large-scale introduction of H2 buses in each city were explored again, and willingness to pay (WTP) for a scenario involving the large-scale introduction of hydrogen buses in the existing city transport system was elicited. Two payment vehicles were used: WTP extra on top of a standard single bus fare, and WTP annual increases in taxes (tax type not specified).

7. Notably, the *ex post* surveys in London, Berlin and Luxembourg collected some data from respondents who had completed the *ex ante* survey one year earlier; these respondents are not included in the analysis presented in the present chapter for reasons of space. Only first-time respondents inter-viewed by random telephone dialling are considered in this chapter. For more details on the other sub-samples see O'Garra (2005b).

8. We consider that comparisons of the bus user samples with the general populations of each city would be inappropriate as there is no reason to assume that bus users have the same characteristics as non-bus users.

9. EPSRC project GR/R50790/01. See, for example, Joffe *et al.* (2004) and Shayegan *et al.* (2006).

10. Refuelling stations were selected through a stratified sampling approach (Malhotra, 1999), where the key stratification variable was average income. London local authority areas were divided into low-income, middle-income and high-income (*Source*: ONS, 2002). Two local authority areas were chosen at random from the highest and lowest income categories, and one from the middle-income category, and commercial refuelling stations in each of these areas were identified and listed. Four stations within each of the strata were selected at random, totalling a sample of 24 existing com-mercial stations. The survey sample was stratified by distance from refu-elling stations (0–100m, 100–200m and 200–500m), in order to test for the impact of distance on opposition likelihood (Dear, 1977).

11. Barnet, Hammersmith and Fulham, Lambeth, Newham and Richmond-upon-Thames.

12. Average annual income values were obtained by multiplying gross weekly household income (£738) by 52 (weeks per year). *Source*: ONS, 2004.

13. Notably, the borough of Redbridge was also included in the study, but cannot be considered a control in relation to Havering, as H2 buses oper-ated in the borough for six months, as part of the wider Clean Urban

Transport for Europe (CUTE) project. Results for Redbridge are therefore not presented in this chapter.

14. Using a Logit regression on the dummy variable H2KNOW (where 1=has heard about H2 vehicles, 0=hasn't heard about H2 vehicles).

15. This question was not included in the *ex post* surveys, as the emphasis in this set of surveys was on attitudes towards H2 buses.

16. Calculated by taking the derivative at the mean.

17. This interaction variable (Havering resident*survey phase) captures changes in the dependent variable *over time* in the borough of Havering only. Any changes that have occurred over time in both the treatment and control boroughs will be captured by the survey phase indicator.

18. Cost of living indices were obtained from Mercer Consultants. All cost of living indices were divided by the cost of living index for Berlin, which would make Berlin the baseline (cost of living=1). Full details of this adjustment process can be found in O'Garra (2000b).

19. Elicited in the same survey, using a similar scale.

20. The emphasis on *opposition* in this section is for several reasons: firstly, the successful development of new facilities tends to be a function of opposition rather than acceptance. Although the development of a new facility or expansion of an existing one might entail benefits to some individuals, it is typically those in opposition who actually signal their preferences; hence, opposition tends to be the key influence rather than acceptance. This may explain why the literature on facility siting has tended to focus on opposition rather than acceptance. Secondly, the study aims to clarify concerns that the public might oppose local H2 storage, on the basis of safety fears or other issues (e.g. Adamson and Pearson, 2000).

21. Using a Logit regression (in which opposition is modelled as a binary variable, where 1=opposes the development and 0=does not oppose it).

22. This approach is known as a 'difference-in-differences' modelling approach. It is typically used for the evaluation of the impact of policies or other interventions in an area (e.g. Meyer *et al.*, 1995). A multinomial logit regression was used to analyse attitudes towards local H2 storage.

References

K-A. Adamson and P. Pearson, 'Hydrogen and methanol: a comparison of safety, economics, efficiencies and emissions', *Journal of Power Sources*, 86, 1–2 (2000), 548–555.

M. Altmann and C. Graesel, 'The Acceptance of Hydrogen Technologies', Report for the HyWeb (Ottobrun, Germany, Ludwig-Bolkow-Systemtechnik GmBH: 1998).

Chiu, Yi-Chang and Tzeng, Gwo-Hshiung, 'The market acceptance of electric motorcycles in Taiwan experience through a stated preference analysis', *Transportation Research Part D*, 4 (1999) 127–146.

M. Dear, *Not On Our Street: Community Attitudes to Mental Health Care* (London: Pion, 1977).

G. Dinse, *Wasserstofffahrzeuge und ihr Funktionsraum – Eine Analyse der technischen, politisch-rechtlichen und sozialen Dimensionen'*, Technische Universität Berlin (in cooperation with Institut für Mobilitätsforschung (ifmo)), 1999.

G. Dinse, 'Akzeptanz von wasserstoffbetriebenen Fahrzeugen' (Berlin: Institut fur Mobilitatsforschung, 2000).

A.H. Eagly and A. Chaiken, *The Psychology of Attitudes* (Florida: Harcourt College Publishers, 1993).

G.O. Ewing and E. Sarigöllü, 'Car fuel-type choice under travel demand management and economic incentives', *Transportation Research Part D*, 3, 6, (1998) 429–444.

J. Foley, *H2: Driving the Future* (London: IPPR, 2001) p. 50.

G. Gamboa and G. Munda, 'The problem of windfarm location: A social multi-criteria evaluation framework', *Energy Policy* (2006) in press, Corrected Proof available online.

G. Gaskell and M.W. Bauer, *Biotechnology 1996–2000: The Years of Controversy* (London: Science Museum, 2001).

K. Haraldsson, A. Folkesson, M. Saxe and P. Alvfors, 'A first report on the attitude towards hydrogen fuel cell buses in Stockholm', *International Journal of Hydrogen Energy*, 31 (2006) 317–325.

W.E. Huffman, M. Rousu, J.F. Shogren and A. Tegene, 'The effects of prior beliefs and learning on consumers' acceptance of genetically modified foods', *Journal of Economic Behavior and Organization*, 63, 1 (2007) 193–206. Corrected Proof, Available online.

IPCC, *Climate change 2001: Synthesis Report*. A contribution of Working Groups I, II and III to the Third Assessment Report of the Intergovernmental Panel on Climate Change (IPCC) (2001).

D. Joffe, D. Hart and A. Bauen, 'Modelling of hydrogen infrastructure for vehicle refuelling in London', *Journal of Power Sources*, 131, 1–2 (2004) 13–22.

M. Karlstrom, 'Local environmental benefits of fuel cell buses – a case study', *Journal of Cleaner Production*, 13, 7 (2005), 679–685.

D.J. Lober and D.P. Green, 'NIMBY or NIABY: a Logit model of opposition to solid waste disposal facility siting', *Journal of Environmental Management*, 40 (1994) 33–50.

London Area Transport Survey (LATS) (2001), Transport for London, London.

U. Lossen, M. Armbruster, S. Horn, P. Kraus and K. Schich (2003), 'Einflussfaktoren auf den Markterfolg von wasserstoffbetriebenen Fahrzeugen' (Factors influencing the market success of vehicles powered by hydrogen), expert verlag. Unpublished.

M.H. Maack, K.D. Nielsen, H.T. Torfason, S.O. Sverrisson and K. Benediktsson, (2004) 'Assessment and evaluation of socio-economic factors', Deliverable No. 12 of Ecological City Transport System (ECTOS) Project (http://www.newenergy.is) (Contract EVK-CT-2000-00033) Icelandic New Energy, Reykjavic, Iceland.

M.H. Maack and J.B. Skulason, 'Implementing the hydrogen economy', *Journal of Cleaner Production*, 14 (2006) 52–64.

N.K. Malhotra and D.F. Birks, *Marketing Research: An Applied Approach*, 2nd European Edition (Harlow: Prentice Hall, 2003).

B.D. Meyer, K. Viscusi and D.L. Durbin, 'Workers' compensation and injury duration: evidence from a natural experiment', *The Amercian Economic Review*, 85, 3 (1995) 322–340.

S. Mourato, B. Saynor and D. Hart, 'Greening London's black cabs: A study of driver preferences for fuel cell taxis', *Energy Policy*, 32 (2004) 685–695.

T. O'Garra, 'Public Acceptability of and Preferences for Hydrogen Fuel Cell Buses and Refuelling Infrastructure', PhD Thesis (2005a) University of London: London.

T. O'Garra, 'Comparative Analysis of the Impact of the Hydrogen Bus Trials on Public Awareness, Attitudes and Preferences: a Comparative Study of Four Cities', Final Analysis Report, AcceptH2 project (2005b) (www.accepth2.com), Research funded by the European Commission under Fifth Framework Programme, Contract ENK5-CT-2002 80653.

T. O'Garra, and S. Mourato, 'Public preferences for hydrogen buses: comparing interval data, OLS and quantile regression approaches', *Environmental and Resource Economics*, 36, 4 (2007) 389–411, Corrected Proof, available online.

T. O'Garra, S. Mourato and P. Pearson, 'Analysing awareness and acceptability of hydrogen vehicles: a London case study', *International Journal of Hydrogen Energy*, 30 (2005) 649–659.

T. O'Garra, S. Mourato, L. Garrity, P. Schmidt, A. Beerenwinkel, M. Altmann, D. Hart, C. Graesel and S. Whitehouse, 'Is the public willing to pay for hydrogen buses? A comparative study of preferences in four cities', *Energy Policy*, 35, 7 (2007) 3630–3642.

M. O'Hare, 'Not on my block you don't: facilities siting and the importance of compensation', *Public Policy*, 25 (1977) 407–458.

ONS, *Region in Figures: London*, Causer, P. and Williams, T. (eds) London: Office for National Statistics, Winter edition, 2002.

ONS, *Region in Figures: London*, Causer, P. and Williams, T. (eds) London: Office for National Statistics, Summer edition, 2004.

P. Schimek, 'Reducing emissions from transit buses', *Regional Science and Urban Economics*, 31 (2001) 433–451.

S. Shayegan, D. Hart, P. Pearson and D. Joffe, 'Analysis of the cost of hydrogen infrastructure for buses in London,' *Journal of Power Sources*, 157, 3 (2006), 862–874.

J. Surrey and C. Huggett, 'Opposition to nuclear power: A review of international experience', *Energy Policy*, 4, 4 (1976) 286–307.

W.A. Watts and W.J. McGuire, 'Persistence of induced opinion change and retention of the inducing message contents', *Journal of Personality and Social Psychology*, 68 (1964) 233–241.

M. Wolsink, 'Wind power and the NIMBY-myth: institutional capacity and the limited significance of public support', *Renewable Energy*, 21 (2000) 49–64.

8

Social Representations of Hydrogen Technologies: a Community-Owned Wind-Hydrogen Project

Fionnguala Sherry-Brennan, Hannah Devine-Wright and Patrick Devine-Wright

Introduction

In order to understand how the possible introduction of hydrogen technologies will be regarded, it is necessary to determine how public understanding and, in particular, risk perceptions are likely to develop. This chapter presents findings about a community-owned wind-hydrogen project (Promoting Unst's Renewable Energy, PURE) in the Shetland Islands of the United Kingdom, which was used as a case study to investigate public understanding of hydrogen within the framework of social representations theory. Interviews with project stakeholders and members of the general public were conducted and a questionnaire containing word associations was distributed to every household on the island to determine whether or not the understanding of hydrogen found in the interview study was reflected in the wider population. Using social representations theory to look at public perception of risk in relation to hydrogen, previous experience and knowledge were identified as important anchors in which to base understanding of hydrogen. Anchoring, combined with objectification of hydrogen in the PURE project, served to increase overall familiarity with hydrogen in order to make it less threatening. Social representations theory was used to go beyond the knowledge-acceptance spectrum to bring together many elements essential in shaping understanding and subsequent evaluation of hydrogen.

Why hydrogen?

Hydrogen has a long history of use in agriculture (e.g. ammonia for pesticides) and in the food industry but in its role as fuel for vehicles or as an energy carrier it remains virtually unknown to many people. The role of hydrogen in the hydrogen economy and its role in the current energy crisis has been the subject of many recent articles and books (for example, see Hoffmann, 2001; Rifkin, 2002). Within the context of the hydrogen economy, the two main uses of hydrogen are for fuel and as an energy carrier or storage medium to produce electricity. For the purposes of this chapter the focus will be on the use of hydrogen as an energy carrier that enables it to be used as a storage medium in conjunction with intermittent renewable energy technologies such as wind turbines. However, as we shall see, the social research literature is not extensive on this topic so consideration of hydrogen for fuel will also enter the discussion.

Concerns for the environment or with climate change have meant that issues such as renewable energy and clean fuel are now entering the public domain and new, unfamiliar technologies with new, unfamiliar risks have to be dealt with. Hydrogen presents both risks and opportunities that have not necessarily been experienced before and technologies associated with hydrogen are rapidly evolving to meet possible future demands for fuel and electricity. Dealing with quickly changing technologies requires significant effort in understanding how they are communicated from the scientific community to the general public.

The role of understanding is critical in the process of risk perception. The social amplification of risk framework (SARF) was developed from research on risk perception of nuclear energy technologies and siting of hazardous facilities (Kasperson *et al.*, 1992; Kasperson *et al.*, 1980). In order to determine influences on public understanding of risk, both scientific and public understanding is taken into account. The SARF has been further enhanced by consideration of risk perception from the perspective of social representations theory (Barnett and Breakwell, 2003; Breakwell, 2001) which adds a more social dimension bringing a broader view of public understanding to risk research. A review of the SARF, social representations and previous risk perception literature on hydrogen is presented later in this discussion, but first we must review some of the key findings emerging from selected recent social research on hydrogen.

Social research on hydrogen

Literature on public understanding and acceptance of hydrogen has mostly focused on the use of hydrogen for fuel e.g. in buses (LBST, 1997; O'Garra, 2005; O'Garra *et al.*, 2005) or taxis (Mourato *et al.*,

2004). Studies carried out in Wales (Cherryman *et al.*, 2005) and by Ricci *et al.* (2006) in South Wales, Teesside, and London, consider the possible uses of hydrogen for electricity as well as fuel.

The introduction of fuel cell buses in various cities around Europe as part of the CUTE project has enabled research on understanding and acceptance of hydrogen to be carried out within an environment that has allowed members of the public to experience hydrogen first hand. Dinse (1999; cited in Schulte *et al.*, 2003) carried out two surveys; first with members of the general public, and second with staff at BMW Germany (Dinse, 2000; cited in Schulte *et al.*, 2003). Both studies looked at knowledge and acceptance of hydrogen with the results being that overall knowledge of hydrogen was low but use of hydrogen in fuel cell vehicles was quite well known and associations with hydrogen were generally positive.

Schulte *et al.* (2003) presented a review of both the LBST (1997) and Dinse (1999, 2000) studies carried out in Germany and, as a result, proposed a general 'model of acceptance'. The model was built on values, wants, and needs and identified the main factors influencing acceptance of hydrogen as: evaluation of risk in scientific terms, public perception of risk, and customer satisfaction. Although differences between scientific and public assessments of risk were mentioned, the means by which these differences arise, or possible consequences of these differences, were not discussed.

The introduction of fuel cell buses in London provided the opportunity for research on levels of knowledge and acceptance of hydrogen in the UK, as described in detail in Chapter 7 and Chapter 10 in this volume. The study carried out by O'Garra *et al.* (2005) looked at levels of knowledge of hydrogen and willingness to pay to use hydrogen fuel cell buses in London. Word associations were used as part of telephone interviews in order to gain an idea of the ways in which people made sense of hydrogen. O'Garra *et al.* (2005) found that 90.3 per cent of respondents were able to provide at least one association with hydrogen, 43.7 per cent provided at least two associations, and 4.8 per cent gave at least three associations. The word associations were categorised into 'positive', 'negative', and 'neutral'. Five positive associations including 'alternative fuel' and 'clean' and eight negative associations e.g. 'bomb' and 'toxic' were found. Total positive associations mentioned were 22 per cent compared to 20 per cent that were negative. The most frequently mentioned associations, however, fell in the 'neutral' category which was split into 'chemical', 'fuel and energy', 'physical properties' and 'other' with chemical associations taking the majority.

The lack of negative associations were used to surmise that public concern with safety of hydrogen was not as great or as widespread as experts supposed, although 60 per cent of participants said they would need more information before making a decision as to whether or not they could support the introduction of hydrogen vehicles, 6.7 per cent of which were specifically interested in safety issues. However, the need for further information identified by O'Garra *et al.* (2005) does not necessarily support the conclusion that safety concerns about hydrogen are not widespread but rather that they are simply unknown or uncertain. This is further supported by the fact that only 43.7 per cent of the sample was able to provide at least two associations with hydrogen implying limited familiarity across the sample.

Conclusions from previous social research on hydrogen can be summarised in the following points:

- the use of hydrogen in vehicles is generally rated in either a neutral or positive light
- safety concerns or risk associations with hydrogen are quite low, but
- there is a need for further information on hydrogen and its associated technologies before further evaluation can be made.

These points illustrate public understanding of hydrogen in what appears to be a rather simplified manner, presuming a positive association between levels of knowledge/education and technology acceptance. The 'deficit model', coined by Wynne (1982) describes how distinctions are often made between objective, scientific, expert knowledge and 'irrational' lay people lacking in the intellectual skills required to fully evaluate a risk. This approach is challenged by the SARF and in the theory of social representations (Moscovici, 1976) which both take into account the social aspects of knowledge in generating meaning rather than the more individualistic information retrieval and processing that is intimated within the deficit model (Joffe, 2003). The usefulness of these approaches is now considered before discussing the case study evidence.

Social amplification of risk framework

In order to explore understanding of risk perception in society, the social amplification of risk framework (SARF) was developed by Kasperson *et al.* (1988). It was proposed as a mechanism through which the expression of risk as a scientific activity and an 'expression of culture' could be understood in an attempt to explain public responses to risk. The social amplification of risk framework is defined as 'the social structures and processes of risk experience ... the resulting repercussions on individual and group perceptions, and the effects of these responses on community, society and economy' (Kasperson *et al.*, 2000).

The means by which social amplification works is as follows: signals are passed from a source, via a transmitter, to a receiver. The source of the information passes it to the transmitter who, in turn, recodes and may amplify the signal before passing it on to the receiver (Kasperson *et al.*, 2000). Transmitters of information are also referred to as 'stations of amplification' because at this stage the incoming signal can be amplified. A signal received through this process is considered to be indirect communication as opposed to direct experience: the processes by which information is transformed when received from a station of amplification may differ from those applied by the individual when information is received directly. There are two processes that can occur during amplification: intensification, where the information is amplified in such a way that the public consider the risk to be of greater impact than expert opinion warrants; and attenuation, whereby the significance of the risk is downplayed so that the public interprets it to be a much lesser threat than expert opinion deems it to be (Barnett and Breakwell, 2003). Information channels that can act as stations of amplification range from media to personal networks and thus incorporate many areas of influence.

The usefulness of the SARF is, however, limited by several factors; Kasperson *et al.*'s (1988) description of the media as, 'homogenous' and the public as, 'passive'; a distinct focus on the individual despite the social aspects of the framework; a separation of the interpretation of novel risk events from pre-existing interpretations of similar issues; the essentially linear process; the negativity of the process of the amplification of risk; and a simplification of the interplay between grounded and mediated knowledge that is called upon when people respond to risks (Petts *et al.*, 2001).

Moreover, Petts *et al.* (2001), in their study on the role of the media in the social amplification of risk issues, suggest that public responses to risk are not driven solely by media coverage. Although the media made extensive reference to popular risk issues – as well as using 'linguistic tags and visual images which resonate with popular fears and anxieties' (p. viii) to describe the new risk – it was suggested that public risk responses were based on:

- patterns of trust in institutions and corporations,
- apparent linkages between risk and non-risk issues and events,
- grounded experience and knowledge including local knowledge and experience derived from neighbourhood and locality,
- baseline mediated knowledge derived from existing patterns of media consumption.

In addition, Petts *et al.* (2001) also demonstrated how the public were not passive receivers of information but actively sought to obtain information about new risks that had not been heard of or experienced before. Individual responses to information were variable due to the interplay of new and existing knowledge, e.g. the effect of previous risk events such as salmonella on the understanding of new risk events such as BSE (Bovine Spongiform Encephalopathy). This is described in terms of 'lay epidemiology' where personal knowledge and experience of previous risk events and health affect how new risks and their possible impacts on health are perceived.

Research into risk perception is often concerned with addressing differences between expert and public understanding of risk issues or events (Barnett and Breakwell, 2003). To take account of this Barnett and Breakwell (2003) build onto the basis of SARF an additional dimension to the amplification process. Along with intensification and attenuation is the process of *representation* – whereby public and expert opinions meet. The processes of representation that are essential in the understanding of public perception of risk are further discussed below.

Connecting risk and social representations

The theory of social representations was developed through a study on psychoanalysis by Moscovici published in 1961 (Moscovici, 1961/1976). The transmission of scientific knowledge, through various forms of communication, to common-sense knowledge is the process that social representations theory aims to explain. Within social representations the two processes of anchoring and objectification help to explain the transformation of knowledge. Anchoring quite simply provides an 'anchor' for something new and unfamiliar. Anchors are usually found in previous knowledge or experience that acts as a library of associations to help make sense of something new. Objectification is the process whereby something abstract and without meaning becomes more concrete. Joffe (2003) likens objectification to symbolisation where meaning is encapsulated within something familiar, a vivid example of this is the swastika. Anchoring and objectification are often identified by the use of metaphors or images used to describe or liken the novel object to something familiar.

Abric (1994) describes how social representations comprise a central core of stable elements arising from a shared value system and provide an organising principle that stabilises the core elements in the process of objectification. The peripheral elements are more dynamic and allow individual representations to exist within a group in which representations are, at the core, consensually shared (Philogène, 2001).

Novel concepts, technologies or events that are unfamiliar may be perceived as threatening or risky, and Breakwell (2001) suggests that the processes of anchoring and objectification in social representations theory could be useful in helping to explain public perception of, and response to, risk. The issue of identity within social representations theory has also been developed in the study of risk. Breakwell (2001) suggests that the use of a particular representation is dependent on how much the representation impacts on identity requirements. Each individual or group experiences different levels of awareness, understanding, acceptance, assimilation, and salience of a particular representation, which leads to varying levels of influence. Principles such as continuity, distinctiveness, self-efficacy, and self-esteem have been identified as important aspects of group identity that influence group belief systems. If a social representation of an object or event is perceived as being risky in terms of having a negative or threatening impact on identity processes then use of the representation will be restricted. It is important therefore to take identity characteristics and social networks into account when discussing the perception of risk events or the change in perception of a risk.

A certain level of risk may be considered synonymous with new technology. Beck (1992) in *The Risk Society*, describes several features of contemporary Western culture that allow heightened levels of anxiety to develop. These include widespread coverage of countless risks by the mass media, lack of trust in experts to protect the individual, lack of know-how of contemporary innovations by experts due to pace of development, and a lack of temporal and spatial boundaries surrounding innovations. It is suggested that new technologies in particular carry a sense of uncertainty that may inherently increase levels of anxiety. However, Washer and Joffe (2006) criticise Beck's model for not suggesting any means of measuring differences in understanding of risk by the scientific and lay communities that would enable empirical examination of, for example, levels of anxiety created by new technological risks. Social representation research on risk has, however, demonstrated that meaning given to a novel technology by a non-expert individual may develop as a protection mechanism that serves to reduce levels of anxiety by creating distance between individual understanding and scientific evaluation of risk (Joffe, 2003).

Revealing the representation

Bauer and Gaskell (1999) provide a succinct and well-defined approach to the study of social representations. Their paradigm for research encompasses four modes of representation (habitual behaviour, individual cognition, informal and formal communication) that interact with each other and with different mediums of representation (e.g. language, images, body movement and sound) to create, maintain, and transform a social representation. These forms of communication are used in a description of communication systems which describes four characteristics, implicit in Moscovici's work, that enable the creation of an operational definition of a social representation:

> the comparison of communication systems in four ways: the content structures, the typified processes of cultivation, the social-psychological functions, and the segmentation of social milieus. (Bauer and Gaskell, 1999, 181)

Thus, by identifying the characteristics of communication systems, and the modes and mediums of representation, the empirical study of social representations is provided with a framework for operation.

Bauer and Gaskell (1999) suggest an 'ideal type' for researching social representations. This consists of seven implications (content and process, multi-method analysis, longitudinal data, social milieus, cultivation studies, crossovers of cultural projects and trajectories, and the disinterested research attitude) that form the ideal paradigm for research but they emphasise that not all are expected to be included in one study.

There are several methods that have been used to look at the structure and content of social representations. The use of word association has been utilised in several studies (for example, see Di Giacomo, 1980; Moodie *et al.*, 1995, Wagner *et al.*, 1996; Markova *et al.*, 1998; Moloney *et al.*, 2005). Word association has a certain characteristic that makes it a useful technique for exploring a representation – that of tapping into habitual or non-reflexive thought. For example, a word association task would ask 'What comes to mind when I say "hydrogen"'? Markova *et al.* (1998) discuss how, when presented with a stimulus word to which response by association is needed quickly, the opportunity to consider or reflect upon a word is reduced resulting in more habitual and, 'culturally shared meaning potentialities' (p. 826). These meaning potentialities are thought to be more stable across a group or culture than meanings associated with more reflexive thought. Word associations also allow the respondent to reply in a less constrained manner, drawing on 'significant categories' (Di Giacomo, 1980) that they may use to anchor the object.

The stability of culturally-shared meanings is linked to the existence of stable core elements that are context-invariant (Wagner *et al.*, 1996), non-negotiable (Moscovici, 1993), and that determine the organisation and structure of the representation (Abric, 1994). Words that, following analysis, do not appear to be stable may be considered as peripheral elements thus allowing for inter-individual variation within a representation. Using word association to determine the structure of a social representation, Wagner *et al.* (1996) illustrated how core elements related to the process of collective symbolic coping – a process used in social representations to cope with threatening objects or events.

Collective symbolic coping

In the process of forming a representation, the unfamiliar is transformed into something that is more familiar. Moscovici (1984) characterises unfamiliarity as the '"not quite rightness" of an object' with which feelings of dread, anxiety or threat may be associated. Rather than allowing the new object to remain threatening it is associated with that which is already familiar i.e. non-threatening. Wagner and Kronberger (2001) refer to this concept as 'symbolic coping' defining it as, 'the process of appropriating the novel and unfamiliar in order to make it intelligible and communicable' (p. 148). The role of the media in communicating novelty can trigger symbolic coping which occurs at both the individual and collective levels as individuals make evaluations based on how the novelty will affect them both as individuals and as the community that they are part of (Wagner *et al.*, 2002). However, the activity of symbolic coping is considered to be collective as individual decisions on naming and categorisation, which form the basis of the processes of anchoring and objectification, are shaped by the framework of shared knowledge that individuals possess as the result of belonging to a particular group (Wagner and Kronberger, 2001).

Within the framework of collective symbolic coping Wagner *et al.* (2002) describe a four-stage process to define an emerging representation. The four stages are:

1. creating awareness
2. production of divergent images
3. convergence of dominant images
4. normalisation

This framework has been used in the case study to help describe the means by which hydrogen is understood by the general public. To illustrate how the stages can be used to identify a social representation, results from the interview study are briefly reviewed here. Within the context of the case study on Unst, the first stage – creating awareness – examines which methods have been used to deliver information about hydrogen and which methods of communication are used by the general public. The second stage – production of divergent images – focuses on functionality of hydrogen and community sustainability that have been derived through various means of communication. The third stage – convergence of dominant images – illustrates how dominant images of community sustainability within broader socio-economic aspects are used to evaluate information provided about hydrogen in the context of Unst. The fourth stage – normalisation – builds on socio-economic aspects and demonstrates how these are used to guide understanding and evaluation of the risks and uses of hydrogen within the context of the PURE project on Unst.

Unst, a windy isle

Unst, the northernmost of the Shetland Islands of the UK (population approx. 600 in 2006) was chosen as a context suitable for study not only because of the installation of a new technology that uses hydrogen as an energy carrier but also because it is a unique social group identifiable by context: a small, remote island, and one that is influenced by a new technology that appears to affect few individuals directly but which has the potential for larger scale influence in the future, both locally and globally.

The community on Unst experienced dramatic change in 1999 when the radar base RAF Saxa Vord, established in 1940, experienced partial closure and, in 2005, the complete closure of the base was announced. The impacts on the community were substantial and included loss of jobs for people who were employed at the base and the loss of almost a third of the population. In response to the 1999 drawdown and sub-sequent closure of the RAF base and local airport the Unst Partnership was established to create and support opportunities for employment through community initiatives that would help to promote and re-generate Unst. Established in November 1999 by the Unst Community Council, the Shetland Islands Council, and Shetland Enterprise Com-pany, the Unst Partnership worked in collaboration with a local en-gineer to develop a community-owned wind-hydrogen project known as the PURE project (Promoting Unst Renewable Energy). With funding secured from siGEN (a large fuel cell company) as well as the European Regional Development Fund (ERDF), Highlands and Islands Enterprise, Shetland Enterprise and the Shetland Islands Council, and technical support from siGEN, Acagen and Robert Gordon's University in Aber-deen, the PURE project won an award for best community initiative from the Scottish Green Energy Awards in 2003. The PURE project was also represented at the International Conference for Renewable Ener-gies held in Bonn, June 2004, which resulted in an International Action Plan in which the PURE project entered a commitment to help to develop community hydrogen opportunities in clean energy solu-tions across the world.

Many people living on Unst utilise diverse means of generating income e.g. small-scale farming (crofting) or running guesthouses, as few employment opportunities exist. There are, however, several businesses and institutions on the island that do offer some employ-ment; these include schools, a retirement home, leisure centre, garage, café, convenience shops, a boat builder, museum, and her-itage centre. One of the problems facing the community is the loss of young people as they move through the education system and are unable to return to the island because of a lack of suitable jobs. Several aspects of the PURE project are especially relevant to the socio-cultural and economic status of Unst not least the provision of employment for at least two graduates from the island, a renewable energy supply for the offices in which the project is based, and the potential for development of a hydrogen research centre in Unst itself with a view to replicate the project within the Shetlands and more widely across Scotland, the UK, and internationally.

Case study: the PURE project

Within the UK there are increasing numbers of community-based renewable energy projects (Walker *et al.*, in press). Combinations of new technologies have created opportunities for communities to create projects to suit their particular needs, a prime example of which is the PURE project on Unst. Although several community-owned wind turbine projects exist around the UK, PURE is unique in utilising both wind and hydrogen to ensure a continuous supply of electricity using hydrogen as a storage medium. In May 2005, the installation of two 15 kW wind turbines, a hydrogen storage system, electrolyser, and a 5kW fuel cell completed the renewable wind-hydrogen system that provides electricity for five units on the small industrial estate where the project is based. The electricity is used to provide heat (for storage heaters), power, and transport. Using the PURE project as a base from which to conduct research about public understanding and risk perception of hydrogen, an interview study was carried out in May 2005 following the official launch of the project and, one year later, questionnaires for further exploration of the social representation of hydrogen were distributed around the island.

Interview study

Interviews were conducted in Unst with 15 members of the general public and four stakeholders in the PURE project to explore their understanding of hydrogen, using the framework of social representations theory (Wagner *et al.*, 2002).

Creating awareness

Sources of information described by participants included newsletters, open days, school visits, and radio interviews by the PURE project employees, as well as newspapers and project reports in the local community council minutes, which are posted on the shop noticeboards. Awareness of the PURE project and its use of hydrogen has been raised through these various sources of information and the ubiquitous social networks. (Italics in quotes is for emphasis; line numbers from interview transcripts are in brackets.)

> ... there's always the shops ... they hae [have] notices up so whenever there's anything on there the notice [goes] up but it's word of mouth is *remarkably* fast around here.
>
> Julian (195–196)

> [Interviewer Question: *how would you find out about what's going on locally?*]: *everybody* knows what's going on locally.
>
> Mark (147–148)

As suggested by Bauer and Gaskell (1999) the variety of modes of communication seen here, a mixture of formal and informal communication, interact with different mediums such as text, images and sound, to provide the basis for the emergence of a social representation of hydrogen. The use of social networks is essential in the creation, development, and transformation of a representation and forms an element of the paradigm for research that Bauer and Gaskell suggest for the study of social representations (Bauer and Gaskell, 1999).

Production of divergent images

The second stage in the emergence of a representation uses the processes of anchoring and objectification to produce divergent images of hydrogen. Metaphors, symbols, and icons can often be identified in anchoring and objectification and help to trace individual experience and knowledge with development of understanding. In the quote from the interviews below it is clear that previous experience, or knowledge, has led to a particular understanding of future developments, like the PURE project, on Unst. Previous examples include the decline of the fishing and boat building industries and the closure of the airport as a result of the transport of offshore oil operations to the mainland Shetland with consequent losses of jobs and people. This is illustrated by these comments:

> I'm quite *interested* in it interested that this development is taking *place* here *but* it's the sorta development that if it is successful it'll not stay here ... you see we have *often* had this problem, anything that is *invented* and developed in the outer isles in a *local* situation it goes to Lerwick [mainland Shetland] ... and the jobs go with it and the *people* goes with it, have to do ... but it means the employment *is* in the peripheral areas rather than in the centre whereas if *this* comes to anything it'll go outta *here* it'll go to somewhere where's it easier made and nearer the source of where it's going to be consumed.
>
> Ray (147–161)

Icons of the PURE project were identified as the two wind turbines (commonly referred to as 'windmills'), a fuel cell/battery hybrid car, and the two main people involved in the project itself. In addition to this, however, understanding of hydrogen was also expressed through the concepts of functionality and community sustainability.

The functionality of hydrogen as part of a wind-hydrogen system was described by interviewees in several different ways. Heat, light, electricity, and transport were all mentioned in relation to hydrogen. These correspond with the 'symbolic resources of everyday life' (Wagner *et al.*, 2002: 324) that are used by people on a daily basis and are commonalities that can be used in communication with others in order to help develop an understanding of hydrogen.

Benefits to the community, particularly in cases where hydrogen was not well understood, were recognised in the form of employment opportunities predominantly reflecting the current socio-economic status of the island. For example, benefits cited included keeping young people employed and on the island, the possibility of self-sufficiency in terms of electricity, or the provision of cheaper electricity for the island.

> ... the fact that we have young graduates on the island already working but the possibility it may be sustained and encourage others, local graduates, to see something that is challenging and innovative on the island that's worth their while to come back for.
>
> June (112–115)

Convergence of dominant images

The third stage in the emergence of a representation implies an evaluation of the information supplied through various sources in order to arrive at a consensual understanding of hydrogen. Similarities between interviewees that were seen at this stage in their representations of hydrogen centred around the issue of community sustainability rather than having drawn conclusions about hydrogen by evaluating its properties, associated technologies, or perceived risk. The underlying social aspects of the social representation of hydrogen on the island are therefore strong as it is deemed important to maintain quality of life on Unst by accepting and supporting the project with which hydrogen is associated. Participants expressed a realistic but positive image of Unst that reinforced consensus on an evaluation of hydrogen necessary to maintain stability of the established way of life,

> I like the *people*, I like (.) the *way* of life, the fact that you make your *own* community ... the climate's horrendous *but* hehe it's *worth* it because if you get one fine day it makes up for a heck of a lot of bad ones'.
>
> Elizabeth (231–235)

Normalisation

The final stage in the emergence of a representation is normalisation. Normalisation demonstrates the functional equivalence of a dominant, consensual belief to scientific knowledge that allows confidence in evaluating a novel object (Wagner *et al.*, 2002) within a broad spectrum of risk and benefit. The dominant belief that can be traced through the previous three stages is that benefit to the community will be conferred by the continuing presence and development of the PURE project and therefore the use of hydrogen within it. This is shown, at this stage, by evaluation of the project in social rather than scientific technical terms.

Questionnaire study – word associations

Following on from the interview study, a questionnaire containing word associations was distributed to every household on Unst in order to see whether or not the understandings of hydrogen found in the interview study were also present in the wider population.

The sample from which results were obtained comprised 48.3 per cent males and 51.7 per cent females with an age range of 81 (mean age 52.4, total sample size = 161). The majority of respondents aligned themselves politically with the Liberal Democrats (36.6 per cent) with Conservatives and Labour being equally represented (8.5 per cent). The main religious group was Christianity (75.6 per cent) of which 34.8 per cent were specifically Church of Scotland. Other religious groups such as Roman Catholic, Humanist and Jewish were also represented but very much in the minority.

Structure of the social representation of hydrogen

Using word associations to look at the structure of the social representation of hydrogen, Table 8.1 gives a comparison between results from this study and those found by O'Garra *et al.* (2005).

The table illustrates the percentage of respondents who associated at least one, two, or three words with the stimulus word 'hydrogen': 88 per cent of the sample was able to produce at least one association with hydrogen, similar to that in O'Garra *et al.*'s study. The percentage of respondents from the Unst sample able to produce at least two words was 78.3 per cent, and 67.1 per cent were able to produce at least three words. The latter figures are markedly different from the sample in O'Garra *et al.*'s study, which suggests greater familiarity with hydrogen or greater personal relevance in the sample from Unst. By

Table 8.1 Comparison between two studies of the numbers of participants who associated at least one, two or three words with the stimulus word 'hydrogen'

Least number of words	Unst (%)	O'Garra *et al.* (2005) (%)
1	88.2	90.3
2	78.3	43.7
3	67.1	4.8

Table 8.2 Examples of words found in the positive, negative, and neutral categories of word associations

Positive	Negative	Neutral
clean	bombs	water
future	expensive	gas
opportunities	danger	chemical
cheap	explosive	light
environmentally friendly	airship	power
jobs		

categorising the words into positive, negative, and neutral categories, 19.7 per cent of first words were positive, 19.7 per cent negative, and 60.6 per cent neutral. Examples of typical words for each category are shown in Table 8.2.

The grouping of words into three basic categories, as in O'Garra *et al.* (2005), allows comparison but loses necessary detail from the data that sheds light on the structure of the representation. For example, the core of a representation comprises elements that are context-invariant (Wagner *et al.*, 1996) and non-negotiable (Moscovici, 1993). By far the most frequently associated words are neutral and can be seen to relate to the chemical or physical properties of hydrogen e.g. gas, element. However, other neutral associations such as power, or wind turbines are dependent on context. Negative associations with hydrogen focus on the combustible nature of hydrogen, e.g. bombs, or its property as a light gas, e.g. airships, both of which are applications of hydrogen that are dependent on the nature of hydrogen itself. Most negative associations are non-negotiable in the sense that hydrogen is a dangerous substance but they are also context-dependent as they are common but not necessary uses of hydrogen. The positive associations made with hydrogen are all context-dependent. Positive associations relate to the properties of hydrogen when utilised in the PURE project e.g. green

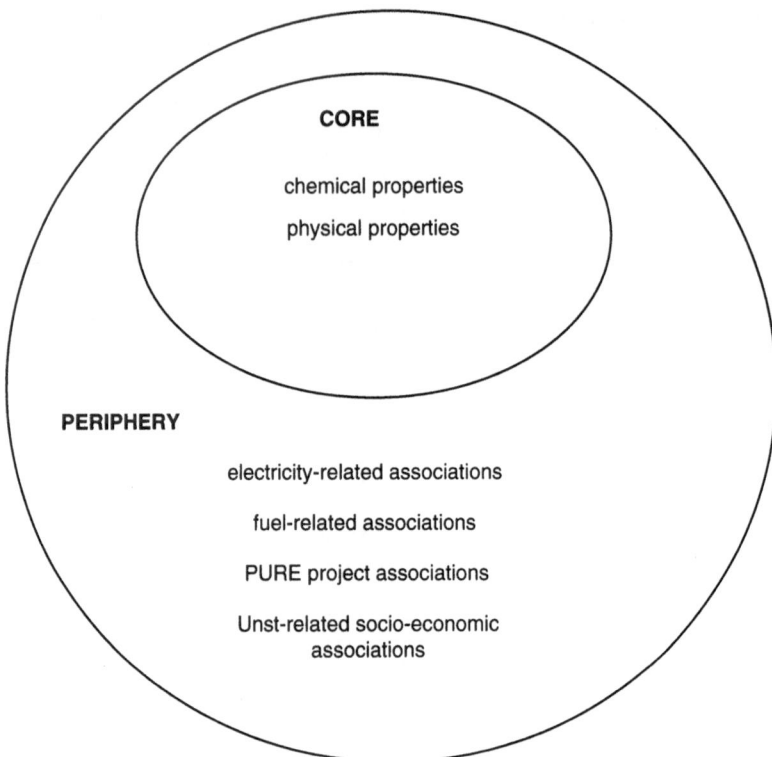

Figure 8.1   Structure of the social representation of hydrogen based on word associations

energy, clean, opportunities. Figure 8.1 illustrates the possible structure of the social representation of hydrogen and the relative positions of the associations within the representation.

Abric (1994) suggests that the central core of a representation is often shared by members of a group and protected from change by the more adaptable peripheral elements which may differ between individuals. Individual variation in the representation of hydrogen is seen by the diversity of terms associated with hydrogen in the periphery of the representation. These also include affective terms that help individuals to understand and evaluate the unfamiliar but are not necessarily consensual (Slovic, 2000). The social representation of hydrogen on Unst reflects the social reality of the context in which it is based. As a result of the context-dependent nature of the representation, salient features of the representation of hydrogen in Unst may not be present in a hydrogen representation for someone living in central London.

Conclusions

The two stages of research helped to identify different aspects of the social representation of hydrogen. In the interview study the process of collective symbolic coping illustrated the importance of all types of social communication networks, from the national press to shop noticeboard. Social communication provided anchors to existing beliefs that were familiar and enabled hydrogen to be objectified and made significant to both individual and community. Generating familiarity in this way helped to make hydrogen non-threatening. Hydrogen was objectified through the PURE project, highlighting the positive impacts on community sustainability rather than focusing on safety or risk concerns. Anchoring and objectification in this case identified aspects of the representation that were used to evaluate hydrogen.

The word association study also helped to describe social representational processes and content by looking at aspects of the representation in more detail. Greater levels of familiarity were found in the Unst sample compared to the London sample (O'Garra *et al.*, 2005) which resulted in lower perceived risk. In Unst the socio-economic (in the positive category) and risk aspects (in the negative category) were seen to be in opposition but, in social representational terms, it is through the process of social debate or argument that a social representation is generated and the process of collective symbolic coping occurs. In the case of Unst, the importance of socio-economic elements, as seen through the need for employment and the provision of businesses generating income for the island, had both personal and community relevance and was thus used to evaluate hydrogen in a positive light. Negative socio-economic impacts resulting from the closure of the RAF base posed a more realistic and imminent threat to the community than explosions of hydrogen, which helps to explain why hydrogen was evaluated in these terms. Both elements, however, remain part of the representation but are differentially activated according to context.

The social representation of hydrogen in London, however, whilst sharing core elements with the representation of hydrogen in Unst (as can be seen by comparison with word associations found by O'Garra *et al.*, 2005) may be less likely to evaluate hydrogen in socio-economic terms as the separation of hydrogen from the individual or group is far greater than for the community in Unst. It is the peripheral elements of a representation that serve to adapt it to context. This emphasises the importance of examining all aspects of the representation to determine how people have come to understand hydrogen and how this might affect their future risk response should hydrogen technologies be introduced on a larger scale in the future. In the case of Unst, social representations theory has looked beyond the 'knowledge-acceptance' spectrum to reveal a complex interplay of elements that influenced understanding and subsequent evaluation of hydrogen.

Acknowledgements

This research was supported by the Engineering and Physical Sciences Research Council (GR/S28082/01), as part of the Supergen FutureNet Consortium.

References

J-C. Abric, 'Central system, peripheral system: their functions and roles in the dynamics of social representations', *Papers on Social Representations*, 2, 2 (1994) 75–78.

J. Barnett and G. Breakwell, 'The social amplification of risk and the hazard sequence: the October 1995 oral contraceptive pill scare', *Health, Risk and Society*, 5, 3 (2003) 301–313.

M.W. Bauer and G. Gaskell, 'Towards a paradigm for the study of social representations', *Journal for the Theory of Social Behaviour*, 29 (1999) 162–186.

U. Beck, *The Risk Society* (London: Sage Publications, 1992).

G. Breakwell, 'Mental models and social representations of hazards: the significance of identity processes', *Journal of Risk Research*, 4, 4 (2001) 341–351.

S. Cherryman, S. King, F.R. Hawkes, R. Dinsdale and D.L. Hawkes, *Public attitudes towards the use of hydrogen energy in Wales* (University of Glamorgan, 2005).

J-P. Di Giacomo, 'Inter-group alliances and rejections within a protest movement', *European Journal of Social Psychology*, 10 (1980) 329–344.

G. Dinse (1999), 'Wasserstofahrzeuge und ihr funktionsraum – eine analyse der technischen, politisch-rechtlichen und sozialen dimensionen. Studienarbeit Technische Universität Berlin in cooperation with Institut für Mobilitätsorschung (IFMO)', in Schulte, I., Hart, D. and van der Vorst, R. 'Issues affecting the acceptance of hydrogen fuel', *International Journal of Hydrogen Energy*, 29, 7 (2003) 677–685.

G. Dinse, (2000), 'Akzeptanz von wassrstoffbetriebenen fahrzeugen – eine studie über die verwendung eines neuen und ungewohnten kraftstoffs. Institut für Mobilitätsorschung (IFMO)', in Schulte, I., Hart, D. and van der Vorst, R. 'Issues affecting the acceptance of hydrogen fuel', *International Journal of Hydrogen Energy*, 29, 7 (2003) 677–685.

P. Hoffmann, *Tomorrow's energy: hydrogen, fuel cells, and the prospects for a cleaner plant* (Cambridge, Massachusetts: The MIT Press, 2001).

H. Joffe, 'Risk: From perception to social representation', *British Journal of Social Psychology*, 42 (2003) 55–73.

R.E. Kasperson, G. Berk *et al.*, 'Public opposition to nuclear-energy – retrospect and Prospect', *Science Technology & Human Values*, 31 (1980) 11–23.

R.E. Kasperson, O. Renn and P. Slovic, 'The social amplification of risk: a conceptual framework', *Risk Analysis*, 8 (1988) 177–187.

R.E. Kasperson, D. Golding *et al.*, 'Social distrust as a factor in siting hazardous facilities and communicating risks', *Journal of Social Issues*, 48, 4 (1992), 161–187.

R.E. Kasperson, O. Renn, P. Slovic, H.S. Brown, J. Emel, R. Goble, J.X. Kasperson and S. Ratick, 'The social amplification of risk: a conceptual framework', in Slovic, P. (ed.) *The Perception of Risk* (London: Earthscan, 2000).

LBST, *The acceptance of hydrogen technologies* – a study carried out by Ludwig-Bolkow-Systemtechnik GmbH (LBST) in co-operation with Ludwig-Maximilians University of Munich (1997).

E. Moodie, I. Markova and J. Pichtova, 'Lay representations of democracy', *Culture and Psychology*, 1 (1995) 423–453.

I. Markova, E. Moodie, R.M. Farr, E. Drozda-Senkowska, F. Eros, J. Plichtova, M.C. Grervais, J. Hoffmannova and O. Mullerova, 'Social representations of the individual', *European Journal of Social Psychology*, 28 (1998) 797–829.

G. Moloney, R. Hall and I. Walker, 'Social representations and themata', *British Journal of Social Psychology*, 444 (2005) 415–441.

S. Mourato, B. Saynor and D. Hart, 'Greening London's black cabs: a study of drivers' preferences for fuel cell taxis', *Energy Policy*, 32, 5 (2004), 685–695.

S.M. Moscovici, *La psychanalyse: son image et son public* (Paris: Presses Universitaires de France, 1961/1976).

S.M. Moscovici, 'The phenomenon of social representations', in Farr, R. and Moscovici, S. (eds) *Social Representations* (pp. 3–69) (Cambridge: Cambridge University Press, 1984).

S.M. Moscovici, Introductory address. *Papers on Social Representations*, 2, 3 (1993) 160–170.

G. Philogène, 'A theory of methods', in Deaux, K. and Philogène, G. (eds) *Representations of the Social* (2001) pp. 39–41 (Oxford: Blackwell).

T. O'Garra, *Comparative analysis of the impact of the hydrogen bus trials on public awareness, attitudes, and preferences: a comparative study of four cities.* Study in the framework of the acceptH2 project: Public acceptance of hydrogen transport technologies (2005). Available at www.accepth2.com

T. O'Garra, S. Mourato and P. Pearson, 'Analysing awareness and acceptability of hydrogen vehicles: a London case study', *International Journal of Hydrogen Energy*, 30, 6 (2005) 649–659.

J. Petts, T. Horlick-Jones, G. Murdock, D. Hargreaves, S. McLachlan and R. Lofstedt, *Social amplification of risk: the media and the public*. Contract Research Report 329/2001 (Sudbury: HSE Books, 2001).

M. Ricci, P. Bellaby and R. Flynn, '"Telling it as it is": typical failings in studies of lay opinion about a Hydrogen Economy', in *16th World Hydrogen Energy Conference*, Lyon, France (2006).

J. Rifkin, *The Hydrogen Economy: the creation of the worldwide energy web and the redistribution of power on earth* (Cambridge: Polity Press, 2002).

I. Schulte, D. Hart and R. van der Vorst, 'Issues affecting the acceptance of hydrogen fuel', *International journal of Hydrogen Energy*, 29, 7 (2003) 677–685.

P. Slovic, *Risk Perception* (London: Earthscan, 2000).

W. Wagner and N. Kronberger, '"Killer Tomatoes!" Collective symbolic coping with Biotechnology', in Deaux, K. and Philogène, G. (eds) *Representations of the Social* (Oxford: Blackwell Publishers, 2001) 147–163.

W. Wagner, N. Kronberger and F. Seifert, 'Collective symbolic coping with new technology: knowledge, images and public discourse', *British Journal of Social Psychology*, 41 (2002) 323–343.

W. Wagner, J. Valencia and F. Elejabarrieta, 'Relevance, discourse and the "hot" stable core of social representations – a structural analysis of word associations', *British Journal of Social Psychology*, 35 (1996) 331–351.

G. Walker, S. Hunter, P. Devine-Wright, B. Evans, S. Hunter and H. Fay (forthcoming 2007), 'Harnessing community energies: explaining and evaluating community-based localism in renewable energy policy in the UK', Special issue of *Global Environmental Politics*, edited by Harriet Bulkeley, 7, 2 (2007), in press.

P. Washer and H. Joffe, 'The "hospital superbug": social representations of MRSA', *Social Science and Medicine*, 63 (2006) 2141–2152.

B. Wynne, *Rationality and Ritual: the Windscale inquiry and nuclear decisions in Britain* (Chalfont St. Giles, Buckinghamshire, UK: The British Society for the History of Science, 1982).

9
Stakeholders' and Publics' Perceptions of Hydrogen Energy Technologies

Miriam Ricci, Paul Bellaby and Rob Flynn

Hydrogen energy is not new science but remains a prospective technology. It is relatively unknown to the public. It might substitute for petroleum and natural gas in powering transport and in heating houses, offices, factories and public buildings. How might potential end-users react to its introduction? How do those who already have a stake in developing the technology at local level envisage its future?

This chapter presents new case study evidence, gathered from fieldwork in three regions of the UK, of views among stakeholders and different 'publics'. We shall show how their 'lay' knowledge is contextual and how the responses to potential hazards, costs and benefits of hydrogen systems vary among them.

Background

With the prospect of rapid depletion of fossil fuels (especially natural gas and oil) and concern about global warming and climate change, governments in the major industrial countries and also those of rapidly developing economies (such as China) have been investigating the potential contribution of alternative energy sources to their energy needs. One possibility is hydrogen.

Hydrogen is frequently described as 'renewable' energy and as 'green' in this and two other senses: that it is 'non-polluting' and that it can be 'democratically' produced. By 'non-polluting' is meant chiefly that, when used, it produces neither greenhouse gases (specifically carbon dioxide, the main man-made contribution to global warming) nor (in significant quantities) the air-polluting nitrous oxides that come from burning oil and natural gas. As a non-polluting fuel it can be used, for instance, in fuel cell and internal combustion-powered vehicles (where the only immediate waste is water vapour), in localised combined heat-and-power systems for buildings and in portable applications (such as durable power for laptops and other electronic devices). Rifkin (2002) claims in addition that the widespread introduction of hydrogen energy technologies would revolutionise the economy and social structure, because it would enable households and local communities not only to use hydrogen but also to produce it from a wide range of locally available resources, and moreover not only to meet their own energy needs but also to contribute any surplus they produce to the national electricity grid – so 'democratising' energy production.

However, hydrogen is not an energy source as such, but an energy *carrier*. Though the most abundant element in the universe, it does not occur in a free form and must be generated using other energy sources. One method of generation is electrolysis, a process in which water is split into hydrogen and oxygen by using electricity from any source, including renewable energy – such as wind, wave and solar (with the aid of photo-voltaic cells), and also nuclear energy. Other processes involve the very high temperatures from some geo-thermal sources or nuclear fission, and otherwise fermentation or gasification of organic matter, such as biomass and coal. At present, hydrogen is usually produced from steam reforming of natural gas, and has many industrial applications as a chemical feedstock but few as energy carrier or fuel. It will be apparent that the pollution-free status of hydrogen depends as much on how it is generated as how it is used, though gains may well be made in using hydrogen to avert local pollution, even when it is produced centrally by means that pollute. The same applies to its status as 'renewable' energy: for instance, generating hydrogen by means of fossil fuels and nuclear power is not renewable. Finally, the 'democracy' of a hydrogen economy would be a function of the extent to which production is distributed rather than centralised. The allure of individual and small community autonomy might well compete with the public interest in thermodynamic efficiency, economic viability and safety.

Among technical experts and energy economists, the principal concerns about the introduction of hydrogen energy are indeed the infrastructure investment costs and relative thermodynamic efficiency of the technology, the possible hazards that hydrogen and its technology present and the consequent safety and recycling issues (see Flynn *et al.*, 2006; Hennicke and Fischedick, 2006; McDowall and Eames, 2006; Ricci *et al.*, 2006a). Major international bodies, energy agencies and private corporations have committed significant investment to research and development of hydrogen energy systems (European Hydrogen and Fuel Cell Technology Platform, 2005; United Nations, 2006). The high-level advantages claimed are reduced greenhouse gas emissions and improved security of energy supply. However, there are still significant uncertainties about the performance and reliability of fuel cell technologies and about the overall feasibility and efficiency of production, distribution and storage systems. Hydrogen's 'risk profile' is substantially different from that of the fuels and energy carriers that are used at present, which requires development of adequate regulatory frameworks.

Last, but by no means least, there are questions about how acceptable the perceived risks, costs and benefits of hydrogen might prove to the public.

Public perception of hydrogen: a review of published studies

Public attitudes towards hydrogen as an alternative fuel and energy carrier have started to become the subject of social research, especially perceptions of the prospective risks to safety associated with hydrogen energy. Findings from previous studies (which are discussed in greater detail in Ricci *et al.*, 2006b, and Ricci, 2006) suggest that public responses to the introduction of hydrogen as a fuel are generally favourable and that safety concerns do not feature prominently. However, the majority of such studies have limitations.

Almost all (Altmann and Graesel, 1998; Altmann *et al.*, 2003; Van den Bosch, 2003; Mourato *et al.*, 2004; Neves and Mourato, 2004; Cherryman *et al.*, 2005; O'Garra, 2005; O'Garra *et al.*, 2005) seem to have been carried out with the purpose of supporting the development of a future hydrogen economy. For instance, studies of public perceptions of public transport powered by hydrogen seek to identify possible barriers to the development and often go on to design means to overcome the barriers. It is also often taken for granted that the predominant public concern will be safety. In a pamphlet published by the think-tank *Demos*, Wilsdon and Willis (2004) advocate an improved approach for engaging citizens in science and technology, arguing that debates over science and technology have too often been dominated by questions of risk assessment and perception, while neglecting more fundamental questions that might be at stake in any technological development: Who owns and controls the technology and why? What are the costs and benefits, and to whom will they accrue? The authors also point out that it is very easy to be carried away with the excitement that surrounds a new technology, and this can sometimes lead to neglect of the untapped potential of the technologies we have at our disposal and overstatement of the benefits of new technologies.

Most studies on public perceptions of hydrogen are based on questionnaire surveys, administered by telephone (as in the EU-funded AcceptH2 project, see Mourato *et al.*, 2004; Neves and Mourato, 2004; O'Garra, 2005; O'Garra *et al.*, 2005) or face-to-face (Van den Bosch, 2003). The study carried out in Wales by a team at Glamorgan University (Cherryman *et al.*, 2005) took instead a qualitative approach. Although quantitative approaches can indeed provide a useful snap-shot of public opinions and statistically representative data, they may do little to help us understand the complexity of how the public approaches unfamiliar technologies, and the social, cultural and economic factors that shape and mediate opinions. Preferences are often investigated by using the 'contingent valuation method', that is by asking people about their willingness to pay a premium for specific goods or services. This method has been criticised (Foster *et al.*, 1997; Kenyon *et al.*, 2003) for its presumption that stated preferences would translate into actual behaviour in real-world situations. Hydrogen is at a very early stage of development and, as is found by all studies, public awareness and knowledge of hydrogen energy are low. Opinions based on inadequate knowledge and experience are likely to be provisional.

The other fundamental assumption that most studies make concerns 'the public', who are usually conceived as generic consumers that make decisions and choices on the basis of 'information' they receive. In the development and diffusion of new technologies, the public is often considered to be the last barrier to overcome, once major technical and economic challenges have been dealt with (e.g. European Hydrogen and Fuel Cell Technology Platform, 2005). The case of hydrogen energy is no exception in this respect: the complexities, uncertainties and disputes surrounding hydrogen futures (highlighted in Ricci, 2006 and McDowall and Eames, 2006) are generally masked under 'iconic', positive images giving the impression of consensus among the experts. Little consideration is given to what role is played in citizens' responses by the way in which the information is presented and how they interpret it, including whether they trust the sources. Moreover, the public itself is generally considered as a homogeneous, uninformed or ill-informed entity that needs to be educated in order to appreciate new technologies. The underlying assumption is that more information dissipates doubts, puts an end to controversies and encourages rational decision-making.

The increasing amount of social research that addresses public perceptions and understanding of various technological issues (radioactive waste management among them, as also genetically-modified organisms (GMOs), nanotechnology, and mobile telecommunications, as shown elsewhere in this book) has led to a more sophisticated concept of 'the public', which recognises the variety, complexity and dynamic nature of public views and concerns about new technologies (Flynn *et al.*, 2006; Groove-White *et al.*, 2000). Irwin (1995) has introduced the concept of 'scientific citizenship' to emphasise the role that public values and concerns about technological developments could play in policy making and risk assessments. Conventional representations of 'the public' neglect the fact that there are different 'publics' with distinctive understandings of scientific issues – a fact that calls for a more flexible and differentiated way of communicating and engaging with citizens.

Again, past research tends to focus only on attitudes towards end-user applications, such as hydrogen fuel cell transport, neglecting the other technology in any 'hydrogen energy system' – the infrastructure of hydrogen production, storage and distribution. Past research has also often overlooked the extent to which hydrogen might entail a significant change in practices, not only by those employed in energy production, distribution and utilisation, but also by those who are end-users in personal transportation and in the home, and by citizens who encounter the technology on the public highway and in their neighbourhood. In short, like any other technology, hydrogen energy has to be interpreted as the core of a complex 'socio-technical system' (Bijker *et al.*, 1987) composed of tangible technological artefacts and less tangible social, political, organisational and cultural components.

Finally, attitudes towards a future hydrogen economy cannot be considered in isolation. If hydrogen were to be introduced as an energy carrier, it would necessarily be part of a wider energy system characterised by a multiplicity of primary energy sources, infrastructures and applications. Therefore, attitudes towards hydrogen have to be placed in the broader context of energy provision and consumption and their environmental implications.

Approach and methodology

This investigation of public perceptions of the emerging 'hydrogen economy' is part of a larger EPSRC-funded research programme on the science and technology of sustainable hydrogen energy (see www.ukshec.org). The study has involved focus groups with members of the general public in three case study areas. Interviews have also been held with key stakeholders (local authority officers and councillors, industrial and commercial executives, regional agency officials) to get an appreciation of how embryonic hydrogen economies are being characterised and developed at local level.

Hydrogen energy technology is not generally understood and so there is a danger that even to provide information may bias how those new to the idea respond. Accordingly, we have been determined to avoid providing all the information that participants might have about it before asking them their views. To avoid this we chose parts of the country in which industrial experience and/or demonstration projects would have given participants some basic knowledge before the focus groups began.

Hydrogen energy technology is not only largely unknown but also little developed on the ground in the UK, but there is funding for renewable energy projects from the EU, matched by UK government. We found that local officials in alliance with local commerce had sometimes sought funding for hydrogen energy development and obtained it. This was the true of Teesside, South Wales and Greater London. The funding in Greater London was to enable participation in a Europe-wide hydrogen fuel cell powered bus project. Each of the 11 participants in the CUTE project was a major city (www.fuel-cell-bus-club.com). The Mayor of London and the London Hydrogen Partnership were promoting a series of initiatives exploring the potential of hydrogen, especially as a means for tackling air pollution in a traffic-congested city. Teesside and South Wales, by contrast to London, were both areas in which the chemical industry had long produced hydrogen on a large scale, not specifically for energy purposes but as feedstock for making other chemical compounds. They were also both areas in which the extractive and manufacturing industries on which the local economy had depended were in decline. Those promoting the hydrogen energy projects in Teesside – for instance the Wind Hydrogen project (www.h2net.org.uk), the Tees Valley Hydrogen Project and the Fuel Cell Application Centre at Wilton, and in South Wales – for instance the Baglan Energy Park (http://www.npt.gov.uk/baglanenergypark), the Hydrogen Valley Initiative, led by the Welsh Development Agency, and the H2 Wales project carried out by the University of Glamorgan, saw them as potential means of regenerating their local economies, and drawing on skills in the local workforces that might otherwise become redundant.

If the fact that hydrogen energy is still relatively unknown to the public justifies our selection of areas of embryonic hydrogen energy development for the study, it also suggests that large-scale survey methods are inappropriate to gauge public perceptions of hydrogen energy and the associated technologies. Participants needed an opportunity to reason about the information we might provide by discussing their views with peers. On the other hand, too few members of the public were likely to have a sufficient stake in a future hydrogen economy to be able to map its benefits, costs and risks in detail and depth, as compared with scientific, policy-making, industrial and commercial stakeholders for a companion study reported in this volume by Eames and McDowall. Accordingly, we avoided prolonged 'deliberative mapping' and decided instead to use focus groups, each meeting for about an hour and half. In total, nine focus groups were conducted, four in Teesside, three in London and two in Wales. We have since distributed further information to these groups and repeated our initial meetings, but the data from this phase is still undergoing analysis and will be reported elsewhere.

Members of the public were recruited through local authorities' public consultation panels; in some areas, there were established 'citizens panels' which were said to be drawn from a representative sample of the local population.[1] The focus groups were for the most part mixed in terms of age, gender, ethnicity and socio-economic group. In London, we sought to compensate for the older age of most recruits there by putting together a group of younger people. Members were specifically recruited who did not have scientific or technological background or close familiarity with energy technologies. Groups varied in size from six to (in one case) 13 members. Meetings were each facilitated by two of the researchers. Discussions were audio-recorded and transcribed in full. Thematic analysis was carried out independently by each researcher, and then validated by 'triangulation' (see Barbour and Kitzinger, 1999; Bloor *et al.*, 2001). Focus groups were first asked about their awareness of general issues about the environment, energy and global warming. Then they were asked to consider different types of energy source, including whether they had heard of hydrogen energy. They were shown some visual representations of hydrogen technology and given brief and simplified explanations.

We also identified and sought interviews with industrial and policy stakeholders in each area. The first contacts were identified in regional policy documents and firms' websites. Additional contacts were made through 'snowball' sampling of individuals or organisations referred on to us by initial interviewees. We conducted mainly face-to-face semi-structured interviews with either individuals or a small group from ten different organisations – including regional agencies, local authorities and industrial companies. In one case only did we conduct an interview by telephone.

In what follows, we present the key findings from the first phase of our fieldwork, organised around salient themes.

Results

1) Why a hydrogen economy? Exploring drivers, expected benefits and constraints

A review of the specialist and popular literature about the hydrogen economy reveals that hydrogen is usually associated with 'clean', carbon-free technologies, beneficial to the environment and linked with improved use of domestic energy sources. This is reflected in the findings from our interviews, which also show further motivations to support a local hydrogen economy.

Stakeholders in regional and local agencies, as well as local authorities, tended to stress the regional benefits of developing hydrogen applications, such as economic regeneration and growth, and job creation. In Teesside, hydrogen was represented as an opportunity to give new life to a declining industrial economy, whilst capitalising on existing skills and infrastructures. This emerged clearly in an interview with a local authority representative:

> *In recent years there has been an awful lot of concern about the future of the chemical industry here and the future of the steel plants. If you take the steel plants and the chemical plants out of Teesside, you're taking away, you know, the economic heart out of the area. There is a lot of high quality jobs involved in this.*

He stressed the combination of factors that make Teesside a unique place to develop a hydrogen economy:

> *You look at what the characteristic of this area is. I mean, we have vast hydrogen storage, underground storage, and a whole network of pipelines and a lot of plants that actually produce hydrogen.*

Later in the interview he added:

> *I want local people in this area to say – there's a future of industry in this area. Erm, you know, a lot of the local kids think – God, you know, there's no future round here. Erm, and so we need to show that there is.*

An economic case for a local hydrogen development was also expressed by stakeholders in South Wales – opportunity to reconfigure an economy in transition by making the most of new technologies:

> *So there must be a huge market potential if we can, you know, really be an early player in the game and so, you know, [this area] happens to be at a time in its existence when structural change in the economy is happening and needs to happen. [This area was] built on agriculture, steel and coal. The steel and coal industry, well the coal industry has gone for the time being at least, the steel industry is a fraction of what it was years ago, agriculture is under pressure, there needs to be a major shift in our economy so we see the new technologies as being part of the way forward potentially. (Local authority representative)*

In contrast, representatives of different Greater London boroughs identified environmental benefits, especially in tackling air pollution, and referred to the Mayor's energy strategy (Mayor of London, 2004) as a driver of hydrogen developments.

On Teesside, a different picture emerged from interviews with industrial stakeholders, which included both producers and users of industrial hydrogen. They looked at hydrogen from an exclusively commercial point of view – as a high value feedstock for industrial uses – and tended to be sceptical about innovative uses of hydrogen as energy. One of them pointed out that virtually all hydrogen produced in Teesside is used in the local chemical industry, so that very little would be left for other applications, especially those of low commercial value.

In the focus groups we introduced the characteristics of hydrogen systems gradually, starting from a discussion about general energy and environmental issues that people were aware of and concerned about. While most people could describe some of the properties of hydrogen (for example, as a gas that is abundant and potentially explosive), knowledge about hydrogen as an energy carrier was limited. Even so, most people considered that hydrogen might meet the need to tackle cogent problems of energy and the environment, such as the increased cost of fossil fuels, their limited availability, air pollution and changes in the global environment and climate. In most cases, comments reflected particular situations that were relevant to people's daily lives, such as increased fuel bills and extreme weather patterns. For example:

> *I think another thing is people always want something that is cheaper. If you can offer this [hydrogen] as a cheap alternative to gas, which is*

rising and rising and rising, you know people are going to be more inter-
ested. (Woman, Teesside)
I think more people are environmentally aware at the moment so that is
why we are still looking for a safer, purer fuel. (Woman, South Wales)
It's said to be clean. (Man, London)
I think recently, you know, with natural disasters that have been hap-
pening, a lot of people have now got this hunch – maybe it could be some-
thing to do with global warming [...] I think people are definitely
concerned about climate change. (Woman, London)

When given more information about hydrogen systems, in particular about how hydrogen might be produced, stored, distributed and used as an energy carrier, participants in our groups expressed a wide range of views. Perceptions of hydrogen were neither entirely positive nor completely negative, but depended upon the local context in which people lived and wider beliefs and values about other energy sources and technologies and broader environmental issues. Participants were able and willing to ask many relevant questions about the 'bigger picture' of the hydrogen economy.

In particular, they wanted to know how hydrogen would be produced, from which primary sources, at what costs and how efficiently, and with what implications for the environment (especially for hydrogen production technologies involving hydrocarbons):

Is it easy to make the hydrogen? (Man, Teesside)
How efficient is electrolysis [...] how much carbon do you actually need
to produce hydrogen from water? (Man, Teesside)
There are lots of uses for the hydrogen once you have got it, but where
do you get it from, how do you produce it and how much does it cost to
produce? (Man, Teesside)
How much fossil fuel to produce it? (Man, Teesside)
For me the basic knowledge I've got of the hydrogen concept, I would
say well, is it cost effective in that as I said earlier about producing it, and
will it relieve the pressure of global warming. (Man, Teesside)

Most people welcomed the idea of getting hydrogen from such renew-able energy as wind power. However, they also realised that it would have serious limitations due to intermittency and seasonality. Industrial stakeholders seemed to share similar concerns. Most of them tended to pinpoint the difficulties in delivering the benefits that hydrogen is usually associated with, as their own livelihood depended upon how hydrogen was produced. For example, during a group interview with representatives of a petrochemical company in Teesside:

> *I've never tried to do this, but I'm sure somebody could, whether it [hydrogen] really is better for the overall environment than the primary fuel being involved. (Supply Chain Director, Petrochemicals)*
>
> *Exactly, if you burn petrol in your car, does that make more of an impact on the, something for global warming, does that make more of an impact than running a steam reformer somewhere, transmitting the hydrogen to a car and then burning it at the car, I'm not sure. (Science Specialist, Petrochemicals)*

Other stakeholders (industrial producers and users) made similar comments:

> *I think there is no benefit from traditional natural gas steam reforming, okay. Because you've got to take natural gas, you've got to convert it into hydrogen. There's no benefit in doing that because you're not using a renewable source of any description. The benefit would be in taking a source like landfill or any [waste] you normally throw away and converting that from carbon to hydrogen. That's where the future from my perspective would be because you're taking something that you would probably throw away, that emits greenhouse gases anyway, you probably cause some sort of environmental damage anyway [...] To use natural gas to make hydrogen doesn't make sense from my point of view. (Facility Manager, Gas Manufacturer)*

Although all industrial stakeholders acknowledged the environmental benefits that hydrogen technologies could deliver in principle, they also indicated that in practice that would entail huge investment costs. Moreover, as one stakeholder pointed out, hydrogen energy's added value might be realised in niche applications, such as in portable technologies, where the real environmental benefits would be minimal:

> *I think the environmental benefits that [my colleague] mentioned are going to come from the transport sector. But I think it's more likely that we're going to see the hydrogen economy working in areas of, how can I ..., where there is some difficulty in terms of energy. So I'm thinking, for example, of portability of fuel cells or maybe running a hydrogen plant in a confined space, or in the middle of nowhere. Am I explaining myself? Cos there is some added value. But that's not where the real benefits that*

we talked about earlier are likely to come from. So for me that's a bit of a confusing situation. (Commercial Director, Ammonia and Fertilisers Manufacturer)

This interviewee was particularly sceptical about claims of developing a future hydrogen economy, as the costs would be impracticable:

I think, if there is a benefit in the hydrogen economy then for me it has to be sustainability. Because we are going to have to do something when our energy, our natural gas and oil reserves disappear, if they do. But erh, it will vary when that's going to be. So sustainability I think is what we are trying to look for. Erm, and I think I feel like added to that point is I cannot see in the near future the economics working with the hydrogen economy and that's the bit which really does make me a little bit cynical. (Commercial Director, Ammonia and Fertilisers)

Similarly, people in the focus groups asked how much hydrogen technologies would cost and how they would compare with conventional and other alternative technologies, in terms of benefits, costs and risks:

Is it going to affect my pocket? That's what 99% of the people would say. (Man, Teesside)
How does it compare with other fuels? (Woman, South Wales)
Is there a safety aspect? (Man, South Wales)
You have got the different health hazard with this one, you've got lots of water vapour coming out [...] if you suffer from asthma, is that going to create a problem? (Man, London)
What are the risks with hydrogen plants and leakage? (Woman, London)

2) Hydrogen energy in practice: stakeholders' and publics' concerns and expectations

Participants in our focus groups were shown visual representations of how a hydrogen economy might look, including images of demonstration projects around the world and in the UK, possible commercial applications for stationary, mobile and portable uses, and storage and distribution technologies, such as hydrogen tanks and refuelling stations. People tended to make sense of hydrogen energy not just as a new (and, for most, unfamiliar) technology, but also in relation to their everyday life practices, such as driving and refuelling their cars, taking a bus and providing heat and electricity in their homes. Some of them were worried that they would not understand the new technology and not be able to refuel their cars as they were used to doing:

> *Is it* [hydrogen] *safe? And is it easy to understand [...] without having a lot of jargon surrounding it? (Woman, Teesside)*
> *Can I be sure that I can top up wherever I go in the United Kingdom? (Man, Teesside)*

In other words, people were interested to understand how their routine activities would be affected should a hydrogen-based economy be realised, and in particular what changes in their behaviour and lifestyles would be required. Such human components of technology, generally missing from technical assessments or simplified as 'public acceptability' issues, were explicitly acknowledged in our focus group discussions.

Among hydrogen proponents there is often an assumption that safety risks will make hydrogen less acceptable to the general public. To support this argument, many commentators (including some of the stakeholders we interviewed) argued that hydrogen's fearful reputation as an explosive gas is enhanced by the Hindenburg accident.[2] Evidence from our focus groups suggests that safety is a concern, but this does not appear to cause outright opposition to hydrogen developments. Moreover, very few people referred to the Hindenburg disaster. Men who had direct experience of the chemical industry were especially likely to voice concerns. They discussed possible hazardous situations and the new technological requirements that might be needed. A man in Teesside illustrated this by recalling the accidents that occurred when natural gas was introduced:

> *It still concerns me though that new technology – although it's an old technology, it's not established worldwide technology – what concerns me is, what a wonderful thing ordinary gas was, but look at the accidents, explosions there was in the early days of that. So it's got to be road tested, so to speak. (Man, Teesside)*

Other male participants were concerned about transferring hydrogen out of a tightly controlled industrial environment into the hands of ordinary citizens:

> *It is like everything, if it is a safe environment, if it is working properly and it is designed like nuclear power stations, like gas power stations,*

whatever, as soon as it goes out of that controlled environment ... To put that in the public forum in a busy rush area, it's like a mini bomb. (Man, Teesside)

Looking at that and from my energy background I'd be worried about safety, [that] would be my primary concern. (Man, South Wales)

What crossed my mind when I saw some of the pictures here, that chap putting the pipe into his car, I was thinking there has got to be some very sophisticated valves involved there, because if there is a leak of gas, you know, then that's pretty volatile stuff isn't it? (Man, South Wales)

In the Young Group in London safety issues were not raised at all. Within other groups possible risks to public safety were discussed in the context of more familiar fuels (such as gas, petrol and (Liquid Petroleum Glass) LPG) and there was a recurring expectation that hydrogen systems, if introduced in the wider consumer market, would be engineered to be safe. In all areas we found that group discussion allowed people with different views about safety to confront their arguments and learn from each other, thereby developing a 'communal' understanding of such complex issues:

I think a lot of people would actually be a little bit frightened of it. (Woman, Teesside)

If you think about it, I mean, it would have to be safe before it was put onto the public market for consumption, wouldn't it. (Man, Teesside)

I would presume that if something got to a stage of being on the road then it would have been tested sufficiently so that it was safe. (Woman, South Wales)

I was unaware of the danger until I came here, I think it is possible that that is the situation with most people, but again I would imagine if it does, if it is used, you, know, the same issues would be covered. (Woman, London)

Most stakeholders indicated that safety is a key factor in developing hydrogen-based technologies, but opinions on how safety issues could be handled in a consumer market varied considerably. Some of our interviewees were lobbying for the development of a local hydrogen economy and this was clearly reflected in their narratives, which tended to highlight the benefits of hydrogen rather than its risks:

But let's get the standards right, let's get the training right and so on rather than thinking how disastrous it could all be. (Regeneration & Renewable Energy stakeholder)

According to most of our interviewees, all the risks that hydrogen energy poses will be 'manageable' and once the public is informed about this, they will accept the new technology. In Wales, for example, the Manager of a local energy agency was confident that any problems would be successfully dealt with:

> *As an engineer I think my fundamental belief is that there is an engineering solution to any problem that you can imagine.*

We found similar optimism among hydrogen supporters in Teesside:

> *There is nothing if you work with it for long enough to know that it's all solvable, it can all be done. (Regeneration & Renewable Energy stakeholder)*

In contrast, industrial stakeholders seemed to have more sceptical views and stressed the contrast between using hydrogen in a controlled environment like chemical complexes and a less controllable consumer environment:

> *Certainly in a petrochemical site hydrogen is considered to be a very hazardous material. You would need to work hard to handle it safely and stop putting it in, you know, a consumer's, an untrained consumer's hands. (Technology Manager, Petrochemicals)*
>
> *That's right. Look what happened to the Hindenburg. (Science Specialist, Petrochemicals)*
>
> *That's the public's perception of hydrogen isn't it. And it's a reasonable one I think you need to be careful. (Supply Chain Director, Petrochemicals)*

In their opinion, using hydrogen as energy carrier and fuel implies venturing out of the 'comfort zone', the rigorous regime of discipline that the chemical industry has put in place:

> *It's a discipline people are used to, the regimes. It's difficult to imagine applying in a consumer environment. (Science Specialist, Petrochemicals)*
>
> *You introduce hydrogen to a regime of circumstance where those things are not normal then you can get into difficulty. If for example, going back to the motor fuel example, you replace a petrol tank with a hydrogen tank we all know how many have messed around with our cars and you know had petrol leaking around the place, well we all have. And we understand that it's dangerous, but to do that with hydrogen without proper*

understanding you're in a totally different ball game. Whereas on a chemical complex you're not in a different ball game from that anywhere else. (Supply Chain Manager, Petrochemicals)

3) The hydrogen economy: a desirable future?

A key finding from our interviews, that is confirmed across other expert accounts (reviewed by Ricci, 2006; McDowell and Eames, 2006) is that hydrogen futures are assessed in very different ways by different stakeholders, so that the overall desirability of hydrogen as an energy carrier becomes problematic. The same can be said about public attitudes towards moving to a hydrogen-based economy.

Despite their initial lack of awareness and detailed knowledge of hydrogen, most participants in our groups were able to question the assumptions that are usually made about the beneficial aspects of hydrogen energy. Attitudes towards hydrogen were shaped by a multiplicity of contextual factors, including but not limited to risk perception. Lay understandings of what a hydrogen economy might look like and mean to people varied along several dimensions, such as the local economic context and labour market experience, the existence of a local industrial infrastructure and its concomitant risks, and the actions of local agencies and hydrogen lobbies.

In Teesside, for example, there emerged a sense of pride about the long-standing industrial heritage of the area and some people thought there would be no better place to develop hydrogen technologies in the UK.

I think the standards are quite good for health and safety [...] I'm a fireman so we do visits to places and you see this area, I would probably say the best area for industry and the way it is run. If anybody could handle it [hydrogen], I would say this area could [...] we started off talking about things like nuclear fuel and things that are associated with hydrogen and if anybody could deal with it, here I think we could here in Teesside. (Man, Teesside)

There's the local knowledge, there's the local expertise and you say the fire department are well trained on the chemical side of things. (Man, Teesside)

However, this was combined with a sort of resignation to living in immediate contact with environmental threats.

We've lived with the threat for so long. (Man, Teesside)
"We are surrounded by it. (Man, Teesside)

> *ICI and all the other companies round here, we've lived with it.*
> *(Woman, Teesside)*

They also felt that if such developments would bring jobs for the local community, then acceptance would increase.

> *Because that would be another way to get this over to the public, this is going to bring X amount of jobs and people would welcome it. (Woman, Teesside)*

Public attitudes also depended on perceived benefits and above all, costs. Across the groups, the recurrent idea was that hydrogen must prove to be cost-effective to consumers, not only to the environment, for people to become interested and support it:

> *It's got to be factual not just promises. I said this a long time ago in this discussion, I said it's got to be assessed and the proof has got to be there. The facts have got to be there to tell us it's a viable concept and all that comes into viable – safety, economy, cost-effectiveness. (Man, Teesside)*

In particular, according to one participant, all innovations must provide an added value to the consumer:

> *They [other innovations] have enhanced the person, haven't they, the mobile phone, all enhances the person individually, how would hydrogen? (Man, Teesside).*

Behavioural changes will chiefly depend on people's ability to afford the new technology:

> *But people can embrace, I find from my experience they can embrace things which are good for the planet, good for your health, and good on every level, beneficial, but if it is not going to disrupt their lifestyle too much, if it is not going to be too costly for them. Because it sounds like in the end, you know, the general public would say how much is that going to cost me, you know. (Woman, London)*

Benefits such as no carbon dioxide (CO_2) emissions and no pollution at the point of use were recognised, especially in the London groups (air quality seems an important concern for Londoners).

In sum, people needed to understand the broader context in which hydrogen energy would be developed and what it would imply not just on a global scale, but locally. Overall attitudes towards hydrogen were mixed and most people, while favouring the uptake of a cleaner fuel, were reluctant to give unconditional endorsement. Although most participants agreed that demonstration projects, showing real applications in action, would increase public acceptance, a few remained sceptical and expressed concerns that other energy-saving technologies that are currently available would lose ground in favour of an uncertain technology such as hydrogen:

> *To me it's just another way, it's just another process in the chain that is going to cost money. It hasn't solved the problem yet. (Man, Teesside)*
> *I personally don't feel that hydrogen is necessarily the Holy Grail that we need tomorrow, there are actually lots of things, much more low cost immediately. (Woman, London)*

4) Issues of trust, responsibility and public engagement

The discussion about hydrogen was intertwined, in all focus groups, with issues of trust in information, institutions, and industrial and governmental stakeholders. Most people thought that a shift to hydrogen energy would require significant commitment and efforts from governments and industries. However trust in those actors was low:

> *The industry ultimately is out to make a profit, the government, well we all know about governments. (Man, Teesside)*
> *They [industry] are not looking to be altruistic in terms of finding the perfect fuel for the world. (Man, Teesside)*

For many participants, central and local government were seen as needing to provide leadership and take responsibility in order to achieve the necessary large-scale and long-term changes required. At the same time, individual consumers were seen as the ultimate drivers. Changes in their behaviour were seen to result from self-interest, linked to direct 'threats' that might be caused by large increases in fuel prices, for example, or financial incentives to move to new technologies:

> *You can't talk to people on the street, they are not interested in electricity or gas and the impacts it's having, they are interested in what the bill is at the end of the month or quarter or whatever it is. (Man, Teesside)*

For a significant minority in each of the focus groups, there was scepticism about whether people would make changes voluntarily, and doubts about whether people would be motivated by the public good.

These findings resonate with those of stakeholders, who stressed the difficulty of realising behavioural change in energy use among the general public. A Fuel Cell stakeholder indicated that there was a mismatch between people's awareness of energy and environmental problems, and their behaviour as energy consumers:

> *Are people more aware that there is a problem? Yes they are and mainly because they are aware of flooding and odd weather patterns. Do they actually have a will to do anything about it, particularly if it involves personal change? No they don't. So if you could come up with something which is cleaner and more stable and more environmentally sound, but doesn't change their way of life then you're on a winner, especially if it's cheaper, but you can't have that because that's Utopia. So I think, I think the transport one is going to be quite a big deal. I think stationary heating power and combined heating power systems is less of a big deal actually [...] As it stands today, I don't think there is any incentive for anybody to do anything unless they're forced to.*

Our focus groups show that people's reluctance to alter their behaviour to achieve a public good – moving to a more sustainable society – is influenced by trust (and distrust) in several ways. First, people feel ineffective as individuals because they do not trust others to modify their behaviour as well. Second, they see no point in taking responsibility if large corporations and governments do not appear to share the obligation.

People also expressed the need to receive sound and reliable information, but could not unanimously identify a source they would trust:

> *I think there is a basic flaw that affects all of us and that is lost confidence in good information coming out [...] so if we have lost confidence in councils, governments, environmentalist groups, people who want to sell you things [...] you can't make wise informed decisions on how to move forward. (Woman, Teesside)*

Towards the end of the focus group meetings, people were asked their views about public participation or engagement in issues around science and technology in general, and the hydrogen economy in particular. Although most expressed support for citizens participating more actively in decisions about technological futures, some questioned the feasibility of participatory processes and how beneficial their outcomes would be:

> *I think people need to be involved, but I think at the end of the day it may not be [that] the most popular decision is the best decision. (Woman, South Wales)*
>
> *Does public involvement not depend upon the type of society we want? I think most of us are used to living in a society in which we are made inactive as citizens by the power of business, of government. We can now foresee a future in which these things are adopted irrespective of public opinion. But can we not also see a future in which they might be adopted through some form of stakeholder and a genuine stakeholder culture? Now that might be more effective as a way of encouraging their adoption, might it not? My guess is actually it wouldn't be, it might actually make things more difficult. However, there might be other benefits to us as human beings, as genuine citizens, which we are not seeing at the moment in the adoption of technological developments. (Man, London)*

All stakeholders felt that there was widespread public distrust of political authorities (especially central government) and industry. Representatives of local authorities and agencies believed their efforts to engage with the local community would help strengthen public trust and build consensus around hydrogen, and more generally, sustainable energy. However, both stakeholders and members of the public felt such issues received little attention at national policy level:

> *I don't think it is more widely disseminated, the hydrogen partnership that you referred to, that yet doesn't have a high profile and needs to develop that profile. I don't sense backing from the UK government for a hydrogen economy. I think there is insufficient investment. (Greater London Borough stakeholder)*

Conclusion

In this chapter, we have used results from both focus groups with the public and semi-structured interviews with stakeholders in three quite diverse areas of embryonic hydrogen energy development in the UK, to suggest that members of the public, policy-makers and industrialists at local level all have much to say of relevance about the future of this technology.

We chose qualitative methods, out of dissatisfaction with the frequent use of sample surveys to elicit views on issues that are surrounded by uncertainty, and especially with polls framed by questions that experts one-sidedly consider critical to the public interest. We sought to ask open, non-leading questions and to provide visual as well as oral and written information, so as not to blind participants with science. This promoted deliberation among participants.

The views of policy-makers and industrialists were more obviously divided by different interests and experience than those of the public. Publics differed hardly at all by individual demographic characteristics, but the wider socio-economic context of their regions had a major impact on their views. The local economic context and labour market experience, the existence (or not) of industrial infrastructure and its concomitant risks, the actions of local agencies and 'marketeers' for hydrogen – all influenced citizens' trust in expert knowledge and their willingness to accept the new technology.

What emerged was far from consensus. Nor was there anything to suggest that what was in effect 'upstream consultation' about this new technology might persuade publics and local stakeholders to accept the new technology. On the other hand, the views expressed were germane to whether hydrogen energy technology eventually becomes a working reality, and they cannot be dismissed as ill informed.

Thus, we would conclude that upstream consultation is not likely to yield easy gains for those proponents of hydrogen energy who seek to use it – it is not to be understood as a way of 'educating' lay people, but it may well ensure informed debate in which different interests and persuasions have much to learn from each other.

Notes

1. As a condition of their informed consent to take part, we undertook not to reveal the names of either stakeholders or members of the public who participated in the study. However, we wish to acknowledge the indispensable support of the London Sustainability Exchange and officials of the local authorities on Teesside and in South Wales, in recruiting for the focus groups and finding places in which to conduct them, and also to thank the participants for so kindly giving their time. The research reported here was funded by the EPSRC (Engineering and Physical Sciences Research Council) through the 'Supergen' programme and UK Sustainable Hydrogen Energy Consortium.
2. The Hindenburg, a German Zeppelin fuelled by hydrogen, caught fire while landing in New Jersey in 1937. As a consequence of the accident, 12 passengers and 23 members of the crew died.

References

M. Altmann and C. Graesel, *The Acceptance of Hydrogen Technologies* (1998) Available online at http://www.HyWeb.de/accepth2

M. Altmann, P. Schmidt, S. Mourato and T. O'Garra, *Analysis and Comparisons of Existing Studies*, AcceptH2 Public Acceptance of Hydrogen Transport Technologies, Work Package 3, Final Report (2003). Available online at http://www.accepth2.com

R.S. Barbour and J. Kitzinger, *Developing Focus Group Research* (London: Sage, 1999).

W.E. Bijker, T.P. Hughes and T. Pinch (eds), *The Social Construction of Technological Systems* (Cambridge, MA: MIT Press, 1987).

M. Bloor, J. Frankland, M. Thomas and K. Ronson, *Focus Groups in Social Research* (London: Sage, 2001).

S. Cherryman, S. King, F.R. Hawkes, R. Dinsdale and D.L. Hawkes, *Public attitudes towards the use of hydrogen energy in Wales*, University of Glamorgan (2005).

I. Christie and L. Jarvis, 'How green are our values?' in Park, A., Curtice, J., Thomson, K., Jarvis, L. and Bromley, C. (eds) *British Social Attitudes: the 18th Report* (London: Sage, 2002).

European Hydrogen and Fuel Cell Technology Platform, *Strategic Research Agenda*, (July 2005). Available online at:_http://www.HFPeurope.org/uploads/677/686/HFP_SRA004_SRA-report-final_22JUL2005.pdf

R. Flynn, P. Bellaby and M. Ricci, 'Risk perception of an emergent technology: the case of hydrogen energy', *Forum for Qualitative Research/FQS*, 7, 1 (2006). Available online at: http://www.qualitative-research.net/fqs/fqs-e/inhalt1-06-e.htm

V. Foster, I.J. Bateman and D. Harley, 'Real and hypothetical willingness to pay for environmental preservation: A non-experimental comparison', *Journal of Agricultural Economics*, 48, 2 (1997) 123–138.

R. Grove-White, P. Machnaghten and B. Wynne, *Wising Up. The public and new technologies*, Centre for the Study of Environmental Change, Institute for Environment, Philosophy and Public Policy, Lancaster University (2000). Available online at http://www.lancs.ac.uk/fss/ieppp/staff/docs/wising_upmacnaghten.pdf

P. Hennicke and M. Fischedick, 'Towards sustainable energy systems: the related role of Hydrogen', *Energy Policy*, 34 (2006) 1260–1270.

A. Irwin, *Citizen Science* (London: Routledge, 1995).

W. Kenyon, C. Nevin and N. Hanley, 'Enhancing environmental decision-making using citizens' juries', *Local Environment*, 8, 2 (2003) 221–232.

Mayor of London, Green light to clean energy The Mayor's Energy Strategy, Greater London Authority (February 2004). Available online at: www.london.gov.uk/mayor/strategies/energy/docs/energy_strategy04.pdf.

W. McDowall and M. Eames, 'Forecasts, scenarios, backcasts and roadmaps to the hydrogen economy: a review of the hydrogen futures literature', *Energy Policy*, 34 (2006) 1236–1250.

S. Mourato, B. Saynor and D. Hart, 'Greening London's black cabs: a study of driver's preferences for fuel cell taxis', *Energy Policy*, 32, 5 (2004) 685–695.

T. Neves and S. Mourato, *Comparative Analysis of Berlin, London, Luxembourg and Perth Ex-Ante Surveys*, AcceptH2, Work package 6, Deliverable 6 (July 2004). Available online at http://www.accepth2.com

T. O'Garra, *Comparative Analysis of the Impact of the Hydrogen Bus Trials on Public Awareness, Attitudes and Preferences: a Comparative Study of Four Cities*, AcceptH2 Full Analysis Report (2005). Available online at http://www.accepth2.com

T. O'Garra, S. Mourato and P. Pearson, 'Analysing awareness and acceptability of hydrogen vehicles: a London case study', *International Journal of Hydrogen Energy*, 30 (2005) 649–659.

M. Ricci, 'Exploring public attitudes towards hydrogen energy: conceptual and methodological challenges', UKSHEC Social Science Working Paper N.13, Institute for Social, Cultural and Policy Research, University of Salford (2006). Available online at: http://www.psi.org.uk/ukshec/publications.htm

M. Ricci, G. Newsholme, P. Bellaby and R. Flynn, 'Hydrogen – too dangerous to base our future upon?' Proceedings of the Hazards XIX ICHEME NW Branch Symposium, 27–30 March, University of Manchester (2006a).

M. Ricci, P. Bellaby and R. Flynn, 'Telling it as it is: typical failings in studies of lay opinion about a Hydrogen Economy', Presented at the 16th World Hydrogen Energy Conference, 13–16 June, Lyon, France (2006b).

J. Rifkin, *The Hydrogen Economy* (New York: Tarcher/Putnam, 2002).

United Nations Environment Programme, *The Hydrogen Economy: a non-technical review*, United Nations, New York (2006). Available online at: www.unep.org

S. Van den Bosch, *Consumer acceptance of a hydrogen fuel cell transport system*, Delft University of Technology, The Netherlands (2003).

J. Wilsdon and R. Willis, *See-Through Science*, Demos, London (2004). Available online at: http://www.demos.co.uk/publications/paddlingupstream

10

Technological Transitions and Public Engagement: Competing Visions of a Hydrogen Fuel Station

Mike Hodson, Simon Marvin and Victoria Simpson

1. Introduction

Over the last five years a new body of literature has attempted to shift away from assessing the impacts of technologies to highlight the multi-actor informed possibilities and constraints for socially shaping systemic technological transitions (TT), involving multiple issues at multiple levels (see for example Geels, 2004; Elzen *et al.*, 2004). Although we are sympathetic to TT approaches and their illumination of the possibilities for broadening participation in managing technological transitions, such approaches say relatively little about the wider role of 'publics' in transitions, the places in which transitions take place and the role of different social interests in shaping the production of societal visions and technological expectations.

Our concerns, and motivation for writing this chapter, are therefore primarily threefold. First, even though TT is concerned with understanding (and with shaping) systemic transitions in the socio-technical organisation of large scale systems and infrastructures the approach has a relatively narrow conception of users that focuses on the users of the approach and policy-makers. The role of publics and wider societal engagement is not systemically considered within an approach that would often require publics' involvement in transitions. Second, the technological transitions approach is largely agnostic about place and scale. While the niche is assumed to be a site of demonstration there is a lack of specificity about where the landscape and regime may be located. At best it is ambiguous about the role of the local, urban and regional, national and international. This is odd in an approach that has to deal with the local demonstration that is then developed – out there – through the system. Finally the approach does not adequately develop an understanding of the power relations and asymmetries in the development of what are claimed to be wider societal visions and socio-technical expectations of technological transitions. Critically we need an understanding of whose visions are normalised, an assessment of the resonances and dissonances between different social interests' visions, and the power of a vision in overcoming resistance and barriers to transition pathways.

But if we are to move the debate forward we want to constructively engage with these issues through the development of an approach through which we can more productively analyse social-technical innovation, publics, place and power. In this chapter we therefore critically assess and positively contribute to wider debates around technological transitions by developing a framework through which public engagement in local contexts can be connected to TT debates. In operationalising this framework, we analyse BP's attempts to locate a hydrogen fuel station in Hornchurch, east London.[1] Here we address the interplay between public engagement, technological development and the local context of its (non-)appropriation. The context within which we address the fuel station case is one in which the dominant discourse of downstream risks and impacts in technology assessment, which has been dominant for many decades, is being challenged from a number of perspectives (see, for e.g. Schot and Rip, 1997; Wynne, 2005; and especially Chapters 2 and 6 in this volume).

This chapter is structured in four sections. Section 2 considers disconnections between publics, place and power in conventional TT and in response develops an alternative framework for analysing technological transitions. Section 3 further develops this framework by operationalising it within and through a case study of socio-technical innovation involving publics, place and power in the development of the Hornchurch hydrogen fuelling station. Section 4 concludes by considering the implications of this approach for TT and by briefly summarising the research implication of this framework.

2. Disconnections of transitions, 'publics', place and power

Technological transitions and managing transitions approaches (Geels, 2002, 2004; Elzen *et al.*, 2004; Rotmans *et al.*, 2001) have developed a multi-actor, multi-factor, multi-level framework for exploring and interrogating socio-technical systems and in understanding the possibilities and constraints on systemic transitions. A key feature of the technological transitions literature is the development of a long-term vision which informs the formulation of short-term objectives and underpins evaluation of existing policy. In transitions approaches, the production of visions is an important participatory process used to engage, inspire and mobilise social actors. As part of a long-term process transition, visions and the goals encapsulated in them are subject to evaluation and modification over time (Rotmans *et al.*, 2001).

There are, however, difficulties with this particular conceptualisation of a vision (see Berkhout *et al.*, 2003). For instance there remain questions about 'who' produces the normative visions underpinning transitions approaches. The processes through which visions are produced requires a focus on 'whose' views inform such visions, and importantly 'who' is excluded, underpinned by what forms of expectations and aspirations as well as resources, through what mechanisms or *fora* were they negotiated, with what forms of dissent and compromise? This, then, relates to what is often seen as a shortcoming of transitions approaches – the motivations, negotiations and unfolding aspects of actors in transitions – even amongst key transitions authors (Rotmans *et al.*, 2001: 15).

Yet transitions approaches, however, do not explicitly say much about the role of 'the public'. As transitions approaches are predicated on multi-actor, multi-factor and multi-level aspects (Elzen *et al.*, 2004) this is undoubtedly an important but challenging issue to address. Often when this gap is noted and flagged-up by transitions researchers, it is in terms of developing the transitions research agenda around understanding of the role of users (Elzen *et al.*, 2004) and moves to 'explicitly incorporate the user side in the analysis' of technological change (Geels, 2004: 897). There has also been acknowledgement of the desirability of a focus on aspects of 'consumption and ways of life' (Elzen *et al.*, 2004: 283). A key figure in TT approaches, Frank Geels (2004: 901), acknowledges that '[t]echno-scientific knowledge has become more distributed over a widening range of actors (universities, laboratories, consultancies, R&D units in firms)', that '[c]ultural appropriation of technologies is part of consumption' (Geels, 2004: 902), but that 'in many studies, markets and users are simply assumed to be "out there"' (Geels, 2004: 902) and that it is therefore necessary that we must pay more attention to interactions between actors (Geels, 2002).

But when we start to think about interactions between actors, the existing notion of 'users' is far too narrow. Questions about who, when, how, on what terms and in what ways publics become involved in transitions are critical to an understanding of systemic innovation and the possibility for fundamental shifts in the ways in which publics conduct their everyday life. TT imply significant shifts in the relationships between technology and society, and consequently the politics of how technological change is managed (Schot and Rip, 1997) in the widest sense of that term. When addressing the challenges and consequences faced by, for example, systemic change in energy, water and mobility systems, Constructive Technology Assessment (CTA) – an approach closely linked to transitions approaches – offers some pointers as to how we might think about the role of publics (see Schot and Rip, 1997; Schot, 2001; Genus, 2006).

Rather than assessing 'black-boxed' technologies in terms of their impacts, CTA broadens out the assessment of technologies to focus on the design of technological developments and the participation of non-technical experts in shaping technological development. This broadening of participants in technological development projects and the aspirations for dialogue should be viewed in the context of wider debates in recent decades around the decline of trust in expert knowledge (e.g. Beck *et al.*, 1994) and with the deficiencies of what has been termed the 'deficit model' (Wynne, 1991) of the process of one-way expert-public science and technology communication predicated on notions of an ignorant public.

The move CTA makes is to bring technology developers together with such interested parties as policy-makers, users and citizens to become involved in the design process. By contrast to traditional technology assessment approaches, which focus on the impact of technologies, there is an important role for human agency across a range of interested parties. In particular there is an emphasis on the anticipations of the future consequences of technologies with the assumption that this encapsulates the values and interests of social actors from various different perspectives. These anticipations may be subject to change as part of an unfolding process and, thus, the reflexivity of various social actors becomes an important facet of CTA approaches, as does the interactions between social actors which underpin this and consequent social learning. Processually, there is an ongoing modulation of demand and supply issues through the interests of different social actors. In this respect there is not just an important contribution in terms of initial phases of design but also through demonstration projects and ongoing processes of social learning.

Taking a focus on an unfolding assessment of technologies through dialogue and interaction, CTA draws on various methods which are not specific to CTA, including: consensus conferences, scenario work-shops, electronic consultation, public inquiries and citizens' juries (Genus, 2006: 14). In short:

> A basic tenet of CTA is that the design of technological develop-ment should be a broader, interactive process including a variety of societal actors in addition to technical experts. The effect of broad-ening the design process is that the designers', users', citizens' and policymakers' ideas and values are articulated quite early, and are negotiated and renegotiated throughout the course of the techno-logy development process (which is itself a process of constant design and redesign). This will counteract the prevalent tendency to organize technology development in a basically linear fashion (from development, to market introduction, to regulation) and will allow for more continuous evaluation and modification of new techno-logies in the making (Schot, 2001: 41).

CTA undoubtedly has aspirations to move on from what Johan Schot suggests is the 'the current patter of technology management ... [which] is to sponsor development and regulate application' (Schot, 2001: 40). There are also significant potential benefits of debate, dia-logue and the development of more socially robust knowledge prior to the development of physical infrastructures and their associated sunk costs. CTA does offer some significant and important insights that could begin to re-populate TT approaches with its missing publics. But there are also four issues raised by CTA researchers that mean we cannot simply and uncritically import this approach into TT.

First, whilst CTA broadens out the possibilities to participate in technology design this usually takes place after the decision to develop a technology has been made and, as such, focuses on the design of technology rather than issues of purpose prior to the busi-ness decision. Second, this begins to make visible hidden aspects of power relationships and structural concerns including issues of access and resources (Genus and Coles, 2005) and, in particular, asks ques-tions about who frames what (in CTA terms) is a focus on technology development? Third, if there needs to be a focus on who frames an issue there are also necessary concerns about what their expectations are and for what purpose is an issue framed in a particular way. Finally, there is importance in analysing the methods and processes through which these expectations are translated into action and through which attempts are made to engage others in negotiating and renegotiating expectations through processes of engagement and participation.

Consequently we argue here that if we are to develop an approach to TT that takes publics seriously then it is critical that we address the following four issues:

- The role of publics in framing visions of the future through technologically-informed change before they are closed.
- The issue of who frames visions of the future and with what expectations
- The engagement or participation processes and methods in negotiating and renegotiating the future.
- The analysis of the relationship between vision and actuality and the lessons we can draw from this.

3. Competing visions of a hydrogen fuel station: developing and demonstrating a research framework

These four issues are now addressed by developing a framework for analysing the role of publics in framing and translating into action technologically-informed visions of the future. We develop the framework through a case study of the Clean Urban Transport for Europe (CUTE) initiative and a related hydrogen fuel station development in Hornchurch, east London (see also Mumford, 2006).[2]

Framing a vision of the future

CTA highlights the critical importance of human agency, contingency and the possibilities of socially shaping technologies (Williams and Edge, 2000 [1996]; Mackenzie and Wajcman, 1999). But by focusing on the design of technologies after the business or policy decision has been taken to develop them, CTA then frames (see Goffman, 1974) technology development as 'designing in action' rather than addressing issues of the purpose of technological development (Wynne, 2005). Our argument is that in view of the challenges to the legitimacy of expert knowledge and the increasingly porous boundaries of scientific and technological knowledge production, the aspirations of CTA, whilst laudable, do not go far enough. Or to be more specific, they do not go far enough 'upstream' (see Wilsdon and Willis, 2004).

In questioning the focus of technology assessment on impacts, CTA seemingly moves some way upstream. Yet, through neglecting issues of purpose, it still ultimately focuses on assessing impacts of technologies (see Wynne, 2005). Wynne has pointed out that in terms of what is frequently claimed to be upstream engagement 'this radical apparent potential is compromised by deeper, less manifest cultural assumptions and commitments framing most such initiatives, and that these problematic foundations have yet to be identified, confronted and changed' (Wynne, 2005, 66–67). The key point for us here, if we think about Wynne's argument – which he makes in the context of the upsurge in contemporary 'participatory' initiatives – is in terms of the ways in which CTA encourages dialogue *after* the business or policy decision has been made, which crucially:

> reflects an assumption that the public meanings, or issue definitions, are naturally and properly the sovereign domain of authoritative expert institutions, and that citizens have no capability or proper role in autonomously creating and negotiating such collective, and potentially more diverse, public meanings. (Wynne, 2005: 67)

If one thinks back to the guiding normative visions fundamental to the iterative process of steering transitions, this then suggests that diverse public meanings may be framed-out or closed-down at the earliest upstream stages of the production of visions (Stirling, 2005). This is important as visions are important media in mobilising and shaping expectations and commitment around transitions (see Russell and Williams, 2002: 60).

Visions have been used in the Science and Technology Studies literature to offer prospective views on the form, features, functions and benefits of technologies in relation to domains of application. In this sense, visions articulated at an early stage of development can be viewed as highly aspirational and be seen largely in terms of their symbolic representational articulation of a future rather than a material one (although this is not to neglect the material production and media of communication of the vision). In this respect visions are 'culturally anchored' (Borup *et al.*, 2006) and offer particular characterisations of the future from the present, often invoking particular attributions of the past. These visions and the goals they outline provide a reference point through which networks can be built, gaining commitments to participate, orientating the actions of potential participants and constituencies, and in persuading potential participants of the desirability of transition (see Russell and Williams, 2002: 60–1). Although visions are not fixed and will change over time with the variety of social interests which become involved, the key point is that there is an issue of whether visions are initially articulated around narrow self-interests rather than in terms of a broader sense of societal purpose. There is, thus, a crucial issue of who, or which social interests, produce these early visions of the future and with what expectations.

To take the example of the CUTE initiative, the vision of the future initially developed was that from 2001 a 'public-private partnership' of multinational corporations and supranational political interests would develop a project in which there would be demonstrations, over two years, of 27 fuel cell powered buses in nine European cities (Amsterdam, Barcelona, Hamburg, London, Luxembourg, Madrid, Porto, Stockholm and Stuttgart). Though the project was a two-year initiative, this needs to be couched in the evaluation of the buses being part of a long-term multinational capital and supranational political vision of the future as encompassing some sort of shift to an alternative fuel and transportation system.

The objectives of the vision were:

- 'To illustrate the large spectrum of different operating conditions [for fuel cell buses] to be found in Europe',
- To assess the 'design, construction and operation of the necessary infrastructure for hydrogen production and refuelling stations',
- There was a focus on the: 'collection of findings concerning safety, standardisation and operating behaviour of production for mobile and stationary use, and exchange of experiences including bus operation under differing conditions among the numerous participating companies for replication',
- Further objectives included an: 'ecological, technical and economical analysis of the entire life cycle and comparison with conventional alternatives' and the 'quantification of the abatement of CO_2 at European level and contribution to commitments of Kyoto' as well as 'investigating the acceptance of these vehicles' (European Commission, undated: 2).

The initiative was part-funded by the European Commission (around €21 million of a total of €60 million), through its Directorate-General for Energy and Transport (DG TREN). The remainder of the funding came from the partnership. The network built around the initiative was brought together by Daimler-Chrysler, included a central role for the energy provider BP and to varying degrees 'more than 40 organisations throughout Europe and the rest of the world are now involved in the project' (European Commission, undated: 4). This included local networks of transport providers, energy suppliers, political supporters etc.

The London project, as one of the nine demonstrations, commenced in 2003 and involved a network including Daimler-Chrysler, BP, BOC, Transport for London, London Buses with First Group as the bus operator and the Energy Savings Trust. A key issue in the CUTE project was the relationship between the functioning of the fuel cell buses and associated infrastructure development. This emphasis on configurations of technologies to be tested is captured in the project's representation of the relationship between technologies and local context (see Figure 10.1), where London was seen as a site within which these technologies could be 'dropped-in', 'tested-out' and 'performance data' extracted to inform subsequent iterations of technology development.

The development of an associated infrastructure was critical to the vision of the CUTE project. In this respect there was a key role for BP in addressing hydrogen fuel station development, which was characterised in the vision as:

Figure 10.1   Representing the CUTE project in London

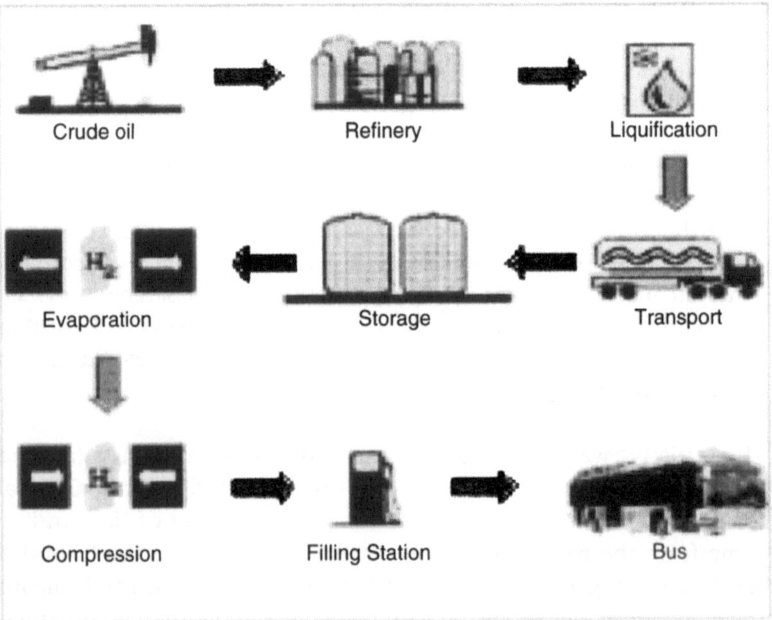

Source: European Commission (Undated)

identifying the most efficient and effective pathways to the Hydrogen Economy. At this stage we don't believe there is one clear winner, so the best way forward is to work a number of these paths by testing various technologies and the customer acceptance of them in detailed ground-level demonstration projects. (BP, 2004)

This was part of BP's 'evolving strategy' of identifying different 'pathways' from a variety of technological options (see Figure 10.2) and then modifying these pathways through feedback from local demonstration projects.

A key aspect of vision of the London demonstration was that there should be a publicly accessible hydrogen fuelling station forecourt, next to an existing petrol filling station in Hornchurch in the east London Borough of Havering. This was the only publicly accessible location of the five fuelling stations being developed across the cities involved in CUTE, and was designed to test out different 'pathways'.

Figure 10.2 Technology 'pathways' in hydrogen energy

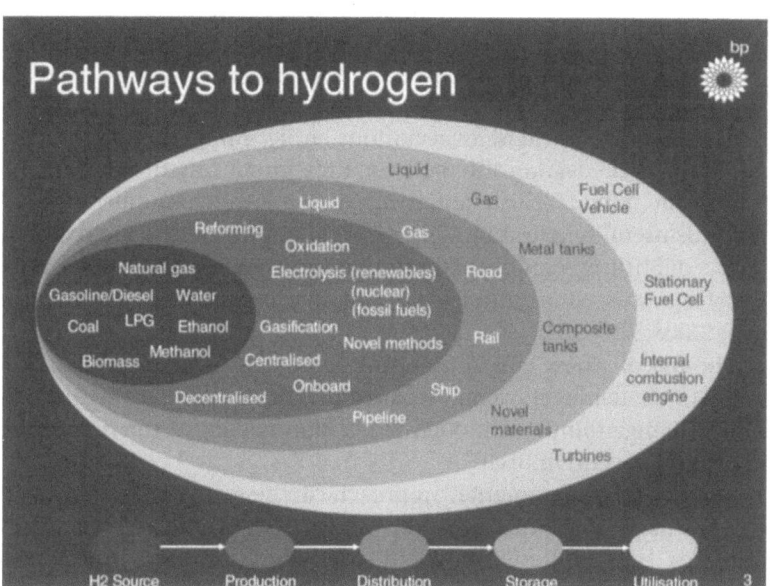

Source: BP

Making-up a vision of the future

The issue this raises is who, or which social interests became involved in producing this vision, with what expectations and with what views of particular publics? The literature in the sociology of expectations (see Borup *et al.*, 2006) offers a fruitful focus here, although we are necessarily selective in drawing on this emerging literature. In the early stages of framing and producing a vision of the future in relation to technological change – given the importance of visions in the subsequent mobilisation and shaping of expectations – the issue becomes one of articulating the variety (or otherwise) of expectations which inform the early stage production of a vision and importantly the ways in which these are communicated.

In focusing on the social construction of visions, through the variety of expectations which inform this, we also acknowledge the differential capabilities and positioning of social interests to meaningfully engage in this process of framing the future. The degree of contestation and the breadth of expectations involved in producing a vision may be narrowly or broadly framed. The importance of whose expectations inform the early stages of a vision are that expectations are constitutive, particularly at the early stages of innovation, in defining roles, attracting interest and building mutually binding obligations (Borup *et al.*, 2006). Additionally, and importantly in view of the spatial shortcomings of transitions approaches, there may also be a significant socio-spatial variability of expectations (Borup *et al.*, 2006), from and of particular places.

Captured within these expectations, either implicitly or explicitly, are views of the relationship between those producing the vision of the future and publics. This relationship can be seen in a number of ways, but it is useful here to highlight views of publics and participation in terms of Stirling's conceptualisation of forms of 'participation' in the social appraisal of technology as normative, instrumental and substantive (2005: 220–222).

The first of these, the 'normative' approach, is underpinned by a sense of the democratic empowerment of citizens participating in decision-making around technological decision-making as 'the right thing to do' and 'an end in itself'.

Second, the 'instrumental' approach can be seen as 'a better way to achieve particular ends' from the strategic viewpoint of incumbent interests, through, for example, the extraction of strategic intelligence from 'participatory' relationships which may also be used in the presentation of particular, already determined, decisions.

Finally, the 'substantive' view focuses on issues of the 'social robustness' of particular technological developments and possibilities in respect of the diverse potential array of social knowledges, values and meanings and the ways in which appraisal is sensitive to differences in this respect and thus produces 'authenticity, robustness and quality in choices that actually result from appraisal' (Stirling, 2005: 222).

In the case of Hornchurch, the vision was produced by multiple actors, with a variety of expectations, at the city-regional, national, European and international levels. The central actors in the initial stages of the project were: the European Union, who co-financed the demonstration; Daimler-Chrysler who developed and manufactured the buses and provided technical support during the trial; and BP who provided the hydrogen-refuelling facilities for the fuel cell buses. There were also roles for BOC who supplied the hydrogen technology to BP in London; Ken Livingstone, the Mayor of London, who backed the introduction of the hydrogen economy in London via emerging planning policies and transport and air quality strategies; Transport for London who were responsible for achieving environmental targets and standards for London's bus fleet as required by the Mayor's Air Quality Strategy; and London Buses Limited who are part of Transport for London and First Group who operate around one sixth of the London bus network. There was additional support from the Energy Savings Trust, through a grant from its New Vehicle Technology Fund Programme (supported by the UK Department for Transport).

The variety of expectations of these actors is captured in Table 10.1. What is also noted in this table is the ways in which these actors made particular attributions, either implicitly or explicitly, to 'publics'.

As Table 10.1 demonstrates, there was a variety of different expectations of the actors involved. Important here was the funding role of the European Commission's DG TREN, the role of networks of multinational capital and the implicit assumptions that hydrogen and fuel cell technologies could be 'dropped-in' to particular 'experimental', 'test-bed' contexts and lessons be learned from these contexts. Cities, in this formulation, were largely seen as 'sites' for technology 'testing' within which context there was limited human agency.

According to a source in DG TREN closely involved in CUTE, 'in the early 2000, the late 90s, [Daimler Chrysler] had a very clear commitment on hydrogen and fuel cells and they thought that it would be a good idea to set up such a project to learn from real life experimentation'. The rationale underpinning this 'real life experimentation', according to a keen observer of the development of this initiative, was

Table 10.1   Actors, expectations and 'publics'

Actor	Expectation	Speaking for 'publics'
Daimler-Chrysler	To be involved in comprehensive fuel cell vehicle test programme on a global scale and to learn from experimentation.	Publics as potential consumers of the hydrogen economy.
European Union	To reduce pollution caused by transport. Understanding of radical social and technical change.	Publics as green consumers in an internationally competitive Europe.
BP	To be at the forefront of the move to a hydrogen economy and to 'test' how the technology 'works' in 'real-world' applications.	Engaging with 'the public' as part of a testing and learning process.
BOC	To lead and develop a programme of initiatives in the evolving hydrogen energy economy.	Publics as potential consumers of the hydrogen economy.
Energy Savings Trust	To support important technological advance in using renewable hydrogen to significantly lower harmful emissions and improve air quality.	Publics as users of new and green technologies.
Mayor of London	Introduced transport and air quality strategies. Supports the development of a hydrogen economy and fuel cell buses. Wants London to be a leading city for sustainable energy.	Improving 'quality of life' issues – air quality, fuel poverty, etc. But also the importance of being seen to be a leading green city in attracting investment.
London Buses and First Group	Hydrogen powered buses to eventually become fully commercialised and replace diesel buses on London streets.	Benefits for 'publics' as public transport passengers, i.e. quieter and more efficient public transport.

'radical social and technical change'. In terms of trying to address this way of understanding large-scale social and technical change the claim was made that multiple fuel cell buses and associated infrastructures needed, in a series of highly 'visible' cities, to be 'tested-out' under a 'variety of conditions'.

A key influence in the development of the vision were the 'big boys' of multinational capital, in that the CUTE initiative, according to an

EU source, 'wouldn't have happened at all were it not for the likes of Daimler-Chrysler, and then, later on the energy companies driving it forwards and putting the whole proposal together ... and then putting out to the cities for interest if you like'. This was because: 'You need the major manufacturers involved to bring this new technology forward or to drive this technology forward'. CUTE addressed not only the functioning of the buses but also the development of a fuelling infrastructure for the buses. For BP the CUTE project was ideal in allowing them to try out several different hydrogen supply methods both small and large scale.

The vision of the CUTE initiative was initially produced though the negotiated expectations of a relatively small group of multinational interests (primarily, Daimler-Chrysler and BP) and supranational political interests (the European Commission), with additional expectations (Mayor of London, Transport for London, etc) informing the development of the project in London.

(Re-)negotiating and materialising a vision through 'participation' and 'engagement' processes

Rather than a neutralised or depoliticised view of processes of participation and engagement, the expectations of particular social interests and the ways in which they are embodied in a vision of the future frames unfolding processes of the negotiation and renegotiation of the future. What is crucial to this is not only the symbolic construction of the vision and the expectations underpinning the vision but how these aspirations inform and translate materially.

This then requires a focus on understanding the ways in which these expectations were negotiated, or formed the basis for interactions around the CUTE project debate in Hornchurch over time. Time is key, as the 'vision' was an expression of the form, features, functions and benefits of the CUTE initiative in relation to local implementation, at an early stage of the initiative but continued to inform subsequent interactions and negotiations as the initiative encountered controversy. The controversy centred around the development of a hydrogen fuelling station in Hornchurch, driven by BP.

Important here are the formal and informal processes of participation and the methods mobilised. The types of methods that are mobilised, the questions asked, by whom, the timing of their mobilisation in terms of a socio-technical transition and the alignment of social interests and the concomitant resources they can draw upon highlights the politicised extent of 'participatory' methods which are often viewed as de-politicised and neutral. It also highlights, in terms of Stirling's three-fold classification, possibilities to 'open-up' or 'close-down' (Stirling, 2005) processes of socio-technical innovation. In addition, with the upsurge of new 'participatory' methods, alongside the plethora of existing techniques and mechanisms, evaluating the role of participatory (engagement) methods becomes extremely confusing. Indeed what may or may not constitute participation has a long history (see Arnstein, 1969), with key concepts not particularly well-defined even taking into account the fruits of this long history (Rowe and Frewer, 2005). With this background in mind, views of what might constitute 'effective' public participation are not only unclear (Rowe and Frewer, 2004), but require a sensitivity (but not a capitulation) to the local context within which they are mobilised.

In the Hornchurch case, following the CUTE announcement, in March 2001, the initiative subsequently moved into a phase of regulation and site development of the hydrogen fuelling station. In July 2002, other actors became involved including Bovis, an engineering company used by BP to undertake construction work and Ozier, a planning consultant commissioned by BP to process the planning application.

The planning application was submitted to the local authority in September 2002 and subsequently involved numerous other actors, including: councillors in the Planning Committee who considered the application; the Health and Safety Executive whose expertise was called upon to assess and advise the local authority on the risks arising from the presence of a hazardous substance to persons in the vicinity; the Environment Agency was required to assess and advise the local authority upon the risks arising to the environment from the presence of hazardous substances; and London Fire Brigade, offered advice about fire safety and carried out various emergency-planning activities.

Local Hornchurch residents were notified by the council, of the planning application for a hydrogen refuelling station at an existing BP petrol station site, in December 2002. Between this time and May 2003 there was a greater involvement of actors who opposed the development. The main objectors were individual residents, the Emerson Park and Ardley Green Residents Association, local councillors and the local media also provided some critical comment. Interactions were mediated through a mixture of formal letters of complaint to the council and local media and informal conversations between residents at the Residents' Association monthly meetings, held at a local school. The position of the residents was one of unhappiness with what they claimed was BP's lack of communication about the development. According to the Chairman of the Residents Association:

> After we had made a number of objections to it [the fuelling station] and raised a number of concerns, the council officers went back to BP about it. We were never given any feedback, we had to go in and find out for ourselves, we never had a meeting offered and we never saw anybody from BP.

In June 2003 the Planning Committee held its first meeting to discuss the development. BP, local councillors and the Residents' Association (including an expert witness supporting the Residents' Association safety concerns) each stated their cases. After consideration of the issues put forward, in July 2003 the Planning Committee refused BP permission. In response BP mounted a campaign against the Committee's decision and revised their planning application, of which the residents received notice in August 2003. In return, the council received a further 26 letters of complaint and a petition and the Planning Committee refused permission for a second time in September 2003. It was at this point that BP appealed again and the decision was made in November 2003 to hold a Public Inquiry. In response to this, in December 2003, the council received another petition and ten additional letters of complaint. In January 2004 the hydrogen bus services were launched with refuelling at a temporary (non-public) facility.

The Public Inquiry, held in May 2004 over three days, involved representatives from BP, the Planning Committee (including an expert witness from the Planning Committee who provided evidence to oppose the development on the Green Belt issue) and a local resident, 'representing' the residents of Cornwall Close, Surrey Drive and Suffolk Way, gave evidence and were cross-examined. After consideration of the issues in July 2004 the Planning Inspectorate and First Secretary of State approved planning permission on the grounds of 'very special circumstances'. The Planning Inspectorate noted:

> ... residents remain fearful of the hazards and the proposals clearly represent an intrusion of inappropriate development, in the Metropolitan Green Belt ... Set against this, the scheme also provides a rare and valuable opportunity, as part of an EU co-ordinated project, to advance the prospect of reducing CO_2 emissions through the use of hydrogen fuel cell vehicles. The participation in the

project, that the development would allow, has the potential to bring environmental improvements on a worldwide scale and to strengthen the competitiveness of the UK industry in this emerging energy sector. (Grantham, 2004: 15–16).

Over the next 12 months there was considerably more active engagement between BP and local residents than had gone before. Four public meetings were organised by BP, which were held in local schools and hotels, and an open day on-site was held when the site was near completion. The stated purpose of the meetings and open day, according to BP, was to give local residents chance to directly speak with representatives of BP who were there to answer any questions or deal with areas of concern. One senior BP official claimed:

> A lot of the wild rumours could be addressed. We could put people's minds at rest on a number of issues. Some of them were just technically wrong and you could explain why and that what they were frightened of was technically impossible. Other things were just giving face-to-face reassurances that certain things wouldn't happen that people were concerned that we would do.

The hydrogen refuelling site began operation in May 2005. Table 10.2 provides a summary of the 'forms of participation' in the Hornchurch case. A key point, however, was that in the 51 months from the initial CUTE announcement to site operation, the local residents were given the opportunity to meet informally with BP for the first time in the 42nd month and the opportunity for three more public meetings and one open day over the following eight months.

Consequences: from vision to actuality?

A concern with a vision and its consequences requires analysing the objectives encompassed in a vision and its production, the unfolding forms of participation in trying to translate a vision into action, the issues raised by these participatory processes and the extent to which the vision materialised. In some respects the initial vision of the CUTE initiative was one which viewed local context as a site from which 'performance data' could be extracted from the demonstration of hydrogen fuel cell buses and associated infrastructures. This technology-driven vision, and its relative neglect of local context, was perhaps unsurprising given the socio-spatial variation (Borup *et al.*, 2006) encompassed by the coalition of social interests of multinational

Table 10.2 'Forms of participation' in the Hornchurch case

Type	Organisers	Where	Involvement	Purpose
Informal meetings	Residents' Association	Local school hall (monthly)	Local residents (usually 50) Local councillors Invited guests	To discuss local issues of concern and decide action
Petitions	Local residents	Submitted to local authority (two in total)	300/400 signatures	To demonstrate community opposition to the hydrogen station development
Protest letters	Local residents, Local councillors	Submitted to Local Authority (36 letters of complaint)	Local residents Local councillors Residents Association Other concerned/ interested parties	To demonstrate community opposition to the hydrogen station development
Media articles	Local residents, Local councillors	Submitted to local newspaper	Local residents Local councillors Residents association Other concerned/ interested parties	To demonstrate community opposition to the hydrogen station development
Formal meetings	Havering Borough Council Planning Committee	Council offices (2 meetings)	Planning committee members Local councillors BP representatives Residents Association Expert witness for the Residents Association	To hear evidence from interested parties and discuss the planning application

Table 10.2 'Forms of participation' in the Hornchurch case – *continued*

Type	Organisers	Where	Involvement	Purpose
Public inquiry	Planning inspectorate, ODPM	Town hall (lasted 3 days)	Planning inspectorate Planning committee members Expert witness for the Planning committee BP representatives Local residents	To have quasi-judicial hearing and make a decision on the granting of the planning application
Public meetings	BP	Local schools and hotels (4 in total)	Local residents Interested parties	Informal face-to-face discussion to answer questions and provide reassurance
Open day	BP	On the development site (one)	Local residents Residents Association Local councillors	To answer questions and to let local residents see the development

capital and supranational political interests involved in its production and the relative neglect of local social interests. The initial sets of social interests and their expectations which informed the vision, thus, encapsulated in many respects an 'instrumental' (Stirling, 2005) view of the appraisal of technology. This is significant if we situate this within the context of expectations and visions being constitutive in defining roles, attracting interest and building mutually binding obligations (Borup *et al.*, 2006).

This particular framing in terms of technological performance, economic costs and operating conditions encompassed little sense of the participation of publics other than as consumers or customers. Consequently, subsequent forms of participation in (re-)negotiating the vision in action became framed through responses to the initial vision. Local objections to the fuel station were mediated through letters, petitions, informal meetings, the local media and a Public Inquiry. These interventions constituted attempts to open-up (Stirling, 2005) technology appraisal and decision-making processes around the fuel station issue.

There were limits to this, in that as the participation process became formalised through planning processes, the appraisal of the technology became institutionalised around downstream concerns, including, for example, risks, hazards and emergency planning. Thus, the 'rules of the game' had been put in place prior to processes of public participation. The point being that a limited degree of a substantive appraisal of the technology only took place within the (seemingly paradoxical) parameters of an instrumental view of the appraisal of technology.

4. Conclusions

This chapter has developed and demonstrated a framework for analysing the interplay between 'public engagement', technological development and the local context of its (non-)appropriation. This provides a framework that attempts to re-connect Technological Transitions (TT) approaches to the role of publics, the specificity of places and competing visions of socio-technical change. By sensitising TT approaches to these issues we have attempted to develop transition perspectives in the following four ways.

First, we have highlighted the critical role and importance of symbolic and often highly partial visions of the future, in the present, through the development of technological change. Critically we need to understand how such visions are constructed, the degree of inclusivity, the assumptions that underpin their conception of a socio-technical systems and the model of social change that is implied by that vision.

Second, we have demonstrated both the variability and particularity of those social interests whose expectations inform a particular view of the future. Critically we need to understand who is involved in developing visions of technological transitions. Which social interests are involved in the construction of visions, who is excluded – either implicitly or explicitly – from the development of visions, and how socially robust and/or inclusive are such visions as a consequence of the contingency and selectivity of their production?

Third, we have illustrated the consequences of these differences in the ways in which particular social interests, and their domination of power relationships, informs the negotiation and renegotiation of a vision of the future in its translation to practice. Critically this involves carefully tracing the unfolding of visions and expectations as they interact with unanticipated social interests that challenge or question the validity or social robustness of a vision that attempts to speak for a collective societal interest.

Finally, we identified the key lessons that can be learned from processes of moving from Vision to Actuality. The key lesson from a TT perspective was the interplay between the closing down and opening up of technological expectations and the highly limited 'substantive appraisal' of the technology prior to wider public engagement.

An important focus for future research thus becomes developing a better understanding of the interplay of vision, expectations, processes and methods in specific contexts. This call is not to be totally reducible or to capitulate to context but to develop sensitivity to context. In doing this, development of cases in particular contexts is not an end in itself but should be used to inform comparison of patterns, trends and regularities. Through such a programme of work the TT approach could begin to, more carefully and systemically, become sensitised and transformed to a wider politics of publics, place and expectations that recognises and works within differential and asymmetries of power.

Notes

1. The authors gratefully acknowledge the support of both the UK Sustainable Hydrogen Energy Consortium, funded through the UK EPSRC, and the CREATE Acceptance project, funded through the European Commission's Sixth Framework Programme, in undertaking this work.
2. Fieldwork took place, in two phases, between January 2004 and January 2005 and between June and July 2006 and included 18 interviews with local

residents, local, regional, national and supranational policy-makers and officials and industrialists. Additionally use was made of documentation – both in terms of those in the public sphere and some internal organisational/departmental documents made available to us on the basis that content and names were not directly drawn upon – and of a number of web sites. Further observational work and discussions were undertaken at a series of relevant workshops.

References

S. Arnstein, 'A Ladder of Citizen Participation', *Journal American Institute of Planners*, 35 (1969) 215–224.

U. Beck, A. Giddens and S. Lash, *Reflexive Modernisation* (Stanford: Stanford University Press, 1994).

F. Berkhout, A. Smith and A. Stirling, *Socio-Technological Regimes and Transition Contexts*, Working Paper Series (SPRU: University of Sussex, 2003).

M. Borup, N. Brown, K. Konrad and H. Van Lente, 'The Sociology of Expectations in Science and Technology', *Technology Analysis and Strategic Management*, 18:3/4 (2006) 285–298.

BP, 'BP Corporate brochure on hydrogen energy' (2004).

B. Elzen, F. Geels and K. Green (eds), *System Innovation and the Transition to Sustainability: Theory, Evidence and Policy* (Cheltenham: Edward Elgar, 2004).

European Commission, *Clean Urban Transport for Europe General Introduction Brochure* (Brussels: European Commission, undated).

F. Geels, 'Technological transitions as evolutionary reconfiguration processes: a multi-level perspective and a case study', *Research Policy*, 31 (2002) 1257–1274.

F. Geels, 'From sectoral systems of innovation to socio-technical systems: Insights about dynamics and change from sociology and institutional theory', *Research Policy*, 33 (2004) 897–920.

A. Genus, 'Rethinking Constructive Technology Assessment as Democratic, Reflective, Discourse', *Technological Forecasting and Social Change*, 73:1 (2006) 13–26.

A. Genus and A. Coles, 'On Constructive Technology Assessment and Limitations on Public Participation in Technology Assessment', *Technology Analysis and Strategic Management*, 17:4 (2005) 433–443.

E. Goffman, *Frame Analysis: An Essay on the Organization of Experience* (New York: Harper and Row, 1974).

R. Grantham, The Planning Inspectorate, *Report to the First Secretary of State – London Borough of Havering Appeal by BP Oil UK Limited* (The Planning Inspectorate: Bristol, 2004 June).

D. MacKenzie and J. Wajcman (eds), *The Social Shaping of Technology* (Buckingham: Open University Press, 1999).

J. Mumford, 'Improving Risk Communication: strategies for public acceptance of new technology involving high impact low frequency risk', Unpublished PhD Thesis, University of Surrey, Guildford, 2006.

J. Rotmans, R. Kemp and M. van Asselt, 'More Evolution than Revolution', *Foresight*, 3:1 (2001) 1–17.

G. Rowe and L. Frewer, 'Evaluating Public-Participation Exercises: A Research Agenda', *Science Technology Human Values*, 29:4 (2004) 512–556.

G. Rowe and L. Frewer, 'A Typology of Public Engagement Mechanisms', *Science, Technology and Human Values*, 30:2 (2005) 251–290.

S. Russell and R. Williams, 'Social Shaping of Technology: Frameworks, Findings and Implications for Policy ...' in Sorensen, K. and Williams, R. (eds), *Shaping Technology, Guiding Policy*, pp. 37–132 (Cheltenham: Edward Elgar, 2002).

J. Schot and A. Rip, 'The Past and Future of Constructive Technology Assessment', *Technological Change and Social Forecasting*, 54 (1997) 251–268.

J. Schot, 'Towards New Forms of Participatory Technology Development', *Technology Analysis & Strategic Management*, 13:1 (2001) 39–52.

A. Stirling, 'Opening up or closing down? Analysis, participation and power in the social appraisal of technology', in Leach, M., Scoones, I. and Wynne, B. (eds), *Science and Citizens*, pp. 218–236 (London: Zed Books, 2005).

R. Williams and D. Edge, 'The social shaping of technology', in Preece, D., McLoughlin, I. and Dawson, P. (eds), *Technology, Organisations and Innovation: Critical Perspectives on Business and Management*, pp. 545–599 (London: Routledge, 2000 [1996]).

J. Wilsdon and R. Willis, *See-through Science* (London: Demos, 2004).

B. Wynne, 'Knowledges in context', *Science, Technology and Human Values*, 16:1 (1991) 111–121.

B. Wynne, 'Risk as globalizing "democratic" discourse? Framing subjects and citizens', in Leach, M., Scoones, I. and Wynne, B. (eds), *Science and Citizens*, pp. 66–82 (London: Zed Books, 2005).

11
Towards a Sustainable Energy Future: Participatory Foresight and Appraisal as a Response to Managing Uncertainty and Contested Social Values

Malcolm Eames and William McDowall

Introduction

In January 2004, the UK Government's Chief Scientific Advisor Sir David King was reported as describing climate change as 'the most severe problem that we are facing today – more serious even than the threat of terrorism' (King, 2004). In the post 9/11 world it has become routine to characterise western society as increasingly dominated by concerns over risk and security. Given the context of Sir David's comments, on a visit to the US administration, it is clear that he intended to highlight and draw into sharp relief the profound long-term and potentially catastrophic risks of human induced global climate change. Whilst some may take exception to the politically charged nature of the comparison, few would dispute the extent to which the twin concerns of climate change and energy policy are central to addressing the fundamental challenge of sustainable development: of creating an environmentally, socially and economically sustainable society for generations to come.

In this chapter – in the context of risk and the public acceptance of new technologies – we focus on long-term energy policy as a response to the threat of climate change and challenges of sustainable development. Specifically we examine the potential for stakeholder participation in the construction and appraisal of energy futures to promote social learning and support better informed, more robust, and democratic decision-making. These themes and issues are closely connected with discussions in Chapters 2, 3, 6, 9 and 10 in this volume, about the prospects for, and limits on, 'public engagement' in technology policy. However, in contrast with the attention previously given to citizens' or users' perspectives, the primary focus in this chapter is on how major institutional 'stakeholders' approach the appraisal of alternative futures.

Since the oil shocks of the 1970s and 1980s, a wide range of scenario planning and forecasting techniques have been developed and employed by both academic researchers and governmental and corporate decision makers with the aim of improving our capacity to cope with the long time horizons, chronic risks and associated market, political, institutional, technological and environmental uncertainties in the energy arena. However, perceptions of the desirability and sustainability of these energy futures often differ markedly between stakeholders and can become fiercely contested. These disagreements typically turn not simply on narrow questions of technical feasibility or cost, but reflect fundamentally differing social values, interests and expectations.

Combining participatory scenario building and multi-criteria appraisal techniques potentially provides a means of 'opening up' social dialogue over the sustainability of contested energy futures. This chapter examines some recent developments in the use of such 'hybrid' methodologies, with particular reference to the case of hydrogen energy.

'Opening up' in energy futures

It is something of a truism to point out that the future is always, to a greater or lesser extent, uncertain. With respect to energy, systems and markets may appear relatively stable and well characterised over the short term. The long capital life of much generating and refining capacity, coupled with extended planning and construction times for new infrastructure all contribute to a system with a high degree of inertia. Hence modellers routinely make predictions and forecasts of future energy prices, demand and resource use, emissions and so on, based on apparently well characterised relationships between economic activity and energy use.

As history has shown however, events frequently intervene in a manner which is difficult to predict. The oil shocks of the 1970s and 1980s, the first Gulf War, 9/11 and the war on terror, all had dramatic impacts on global energy prices. More broadly uncertainties over geopolitical, economic, technological and environmental changes all render long term forecasts deeply problematic.

The energy models and forecasts produced and used by governments and international organisations typically explore some of these sources of uncertainty, such as rates of economic growth and technological change, but play down others. For example, the annual projections of the International Energy Agency, US Department of Energy, and forecasts from the UK DTI and the European Commission all consider fossil fuel availability and price long into the future (IEA 2004; DTI 2000; EC DG Research 2003). Rarely, however, do they systematically include consideration of the potential impact of global conflict on oil and gas prices, preferring to focus on 'market fundamentals' of estimated resource, refining capacity, projected demand and so on. This is despite the very clear impact of geopolitics on real oil and gas prices over the decades.

Furthermore, many such forecasts tend to present the future as a largely predetermined extension of current trends. This can serve to disguise the degree of societal choice in long term energy planning, by presenting future demand patterns, for example, as a knowable variable, rather than as the result of strategic investments, policy decisions or the social practices and day-to-day choices of individual consumers.

By contrast scenario planning techniques seek to recognise and respond to systematic uncertainty by 'opening up' consideration of a wider range of possible futures.[1] Scenario planning first came to prominence in the energy arena following the oil shocks of the 1970s and 1980s. Exploratory scenarios, in particular, have been widely used by business and governmental organisations to support the development of long-term strategic planning and policy making (Shell International, 2001; Foresight Programme, 2001; Nakicenovic and Swart, 2000). Rather than extrapolating from existing trends, exploratory scenarios seek to inform strategic action by illuminating underlying drivers of change, often drawing upon tacit knowledge and stakeholder expertise, to build internally consistent storylines describing a range of plausible futures. Exploratory scenarios are particularly suited to helping decision makers handle risk and uncertainty by testing the robustness of policies, plans and technologies against a range of possible futures.

However, the future is not simply a neutral space waiting to be revealed to us. Predictions, forecasts and scenario studies alike can all play a performative role, mobilising resources, and shaping action and expectations. Moreover, the future is often a deeply contested arena in which different normative visions compete, reflecting the ideological positions of their proponents (Brown *et al.*, 2000). 'Futures studies', consciously and unconsciously, reflect and advance particular world views.

Unlike many scenarios and foresighting techniques, 'backcasting' is an explicitly normative approach which recognises and sets out to exploit this performative function. Backcasting emphasises the role of human agency in delivering desirable futures, rather than on adapting to anticipated futures outside of human control (Robinson, 1990). Backcasting involves developing a vision of a desirable future, and working back from that future to the present in order to understand the steps that must be taken to get there.

It is perhaps not surprising, then, that the backcasting approach has been closely associated with recent work from the Netherlands and elsewhere on the concept of managing 'socio-technical transitions' towards sustainability. Transition management approaches emphasise the role of future visions in guiding technological transitions towards sustainability (Kemp *et al.*, 1998; Berkhout *et al.*, 2004), and focus on 'systems innovations' rather than the diffusion of individual 'environmentally friendly products' (Berkhout, 2002). The use of backcasting in this perspective draws on evidence from studies of the dynamics of expectations in technological change, which claim that shared visions of desirable futures can be important in achieving the alignment of socio-technical systems necessary for change (Van Lente, 1993).

However, the primacy of the 'vision' in transition management demands critical attention, for both normative democratic and substantive reasons: who has a stake and voice in the future, how their expectations are represented and how competing visions of the future are appraised and evaluated are of critical importance.

This suggests another role for the vision, as a forum for opening up discussion and debate about social priorities with respect to (in this case energy) technology futures. As Berkhout *et al.* (2004, 59) argue:

> the real value of the notion of the 'guiding vision' ... does not lie, as is often implied, in its apparently unproblematic normative policy credentials. Quite the contrary: by focusing on the role of guiding visions, attention is concentrated on the importance of legitimate and effective deliberation and learning, and on the crucial role of providing for plurality, reversibility and sustained dissent.

The development of tools and approaches for participatory engagement with energy futures is essential to realising this role for guiding visions. A number of recent studies combine participatory scenario building and visioning with multi-criteria appraisal, as a way to open up dialogue around energy futures.

The Tyndall Centre for Climate Change Research, in their integrated scenarios project, constructed a series of backcasting scenarios that illustrate how the UK government's target of a 60 per cent cut in UK carbon emissions by 2050 might be achieved. These scenarios were developed through a participatory process, involving workshops to validate the future visions, explore the steps that would be needed to achieve them, and explore their relative sustainability through a participatory multi-criteria appraisal workshop (Anderson *et al.*, 2006). Similarly, Kowalski *et al.* report on the development of backcasting scenarios for the deployment of renewable energy in Austria, and the use of the PROMETHEE multi-criteria appraisal procedure to explore their relative sustainability at different scales (Kowalski *et al.*, 2005).

Such studies share an emphasis on broadening the range of uncertainties considered, exploring divergent perspectives, and making clear the need for difficult social choices as well as technical analysis. The hybrid methodologies used, combining participatory scenarios and multi-criteria appraisal, represent an important methodological development to overcome the shortcomings of simplistic forecasts and 'best choice' approaches to appraisal. We now turn to a more detailed example of the development of a participatory scenario-building and appraisal methodology, and its use in exploring the possibilities of a sustainable hydrogen energy system for the UK.

Participatory appraisal of hydrogen futures: a case study from the UK Sustainable Hydrogen Energy Consortium

Hydrogen has the potential to play a major role in a clean and sustainable energy system. However, much like energy policy in general, hydrogen debates are characterised by widely varying understandings of what a sustainable hydrogen energy system might look like (McDowall and Eames, 2006a). The following case study used participatory approaches to open up the appraisal of prospective hydrogen energy systems. The research was undertaken as part of the work of the UK Sustainable Hydrogen Energy Consortium (UKSHEC) between February 2004 and June 2006.

Research design

An overview of the project structure and methodology is provided in Figure 11.1. Briefly, the project involved the participatory development of a set of visions of a hydrogen economy for the UK. These visions were then refined through consultation, and their relative sustainability was appraised using a 'Multi-Criteria Mapping' approach.

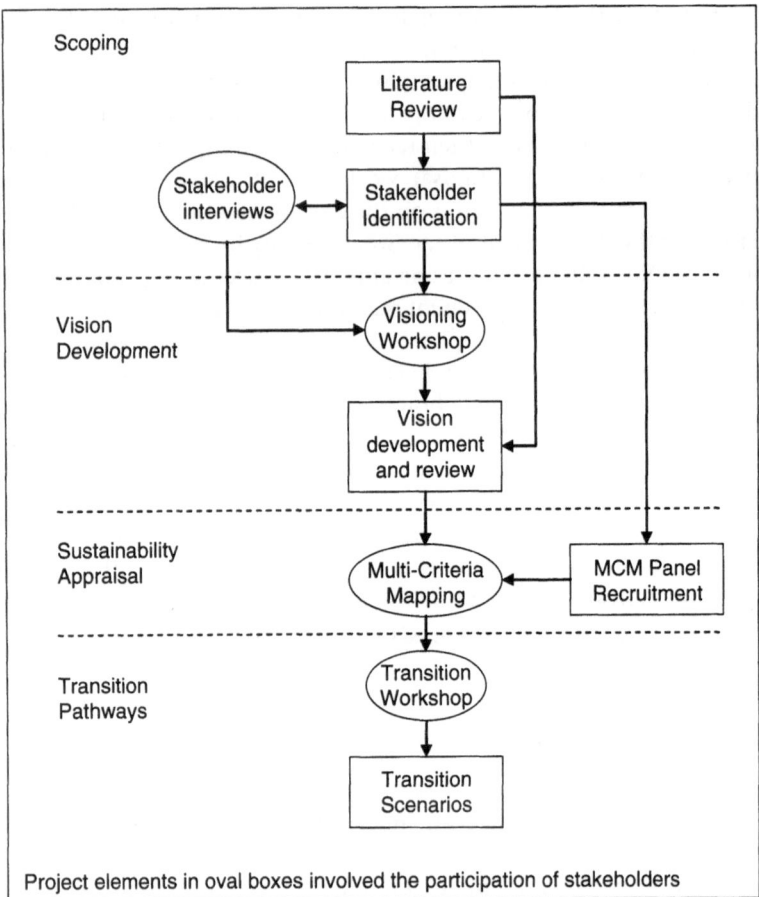

Figure 11.1   Overview of the UKSHEC scenarios project

Finally, a further workshop developed 'transition scenarios', exploring how the visions might be achieved. This case study focuses on the scoping, vision development and sustainability appraisal phases of the work.[2]

Stakeholder involvement

The legitimacy and value of the outcomes of any participatory process are dependent on identification and engagement of appropriate stakeholders, that is, those with either relevant knowledge or an interest in the outcome of the appraisal process. For the purposes of this study, with an emphasis on opening up different perspectives on the long-term appraisal of hydrogen futures, the focus has been to achieve broad transdisciplinary expert participation, rather than to involve lay citizens. While wider publics clearly have an interest in the future of hydrogen, this study aimed to explore a wide range of detailed views for which expert contributions would be most relevant. This is not to deny the importance of public engagement at all stages of technology and futures appraisal, and indeed the methodology could be usefully adapted to engage lay citizens.

Stakeholders were identified through a review of membership of relevant UK steering groups, partnerships and networks associated with hydrogen energy, and through a process of 'mapping' the key areas of the hydrogen production, supply and end-use chains to ensure that participants from all relevant sectors were invited to take part in some form. Scoping interviews with key members of the 'hydrogen community' were also used to elicit views about who should be involved in the process. Efforts were made to ensure that there were sceptical, as well as enthusiastic, viewpoints represented in particular at the appraisal stage. The purposive sampling approach used, which aimed to involve participants from a range of backgrounds, as well as the small number of participants, clearly means that the results from this work cannot be extrapolated to the public at large, or even to the UK hydrogen community as a whole. Rather, the results of the appraisal provide insight into the range of arguments and perspectives, and illustrate the some of the more contentious issues at stake.

Different stages of the study required different levels of stakeholder engagement, commitment and expertise. The 'visioning' workshop used to inform the development of the visions aimed to represent as wide a range of stakeholder opinion as possible, and involved more than 40 participants. The interdisciplinary expert panel which appraised the sustainability of the visions was necessarily much smaller, involving 15 participants from a range of professional and disciplinary backgrounds. Participants took part on the basis of their individual expertise rather than as representatives of their institutions: see Table 11.1.

Constructing the visions

The task of vision development was to capture the wide variety of possible hydrogen futures in a set of credible, transparent and internally consistent end points, without either developing an unmanageable number of visions or making them too vaguely defined to be useful.

Table 11.1 Composition of expert panel

• Nuclear industry expert	• Health and safety regulator
• Carbon trust analyst	• Energy policy researcher
• DTI policy-maker (Department for Trade and Industry)	• Senior oil industry participant
• Fuel cell industry participant	• DfT policy-maker (Department for Transport)
• Sustainable energy consultant	• Automotive industry participant
• Industrial gases industry participant	• Regional government policy-maker
• Energy technology researcher	• Climate scientist
• Environmental campaigner	

In differentiating a set of futures that aim to map out a possibility space, a number of approaches are possible. Previous exercises have frequently focused on the social and economic worlds (and drivers) in which alternative technological systems are thought to be more or less likely to evolve (for example Watson *et al.*, 2004). However, in order to appraise the relative sustainability of the choices facing us with respect to hydrogen, the UKSHEC visions are principally defined in terms of technologies and infrastructures, so that the appraisal will reflect views on technological systems, rather than simply on the desirability of particular social worlds. Given what we know of the co-evolution and co-construction of socio-technical systems, this distinction is potentially problematic, as technologies and social values are inevitably and intimately intertwined. However, as we found, participants often take very different views on the way in which society and technology will co-evolve. Structuring the visions and appraisal in this way therefore allows a rich exploration of these different perspectives. It also allowed us to engage with the detail of the various technologies involved rather than simply treating hydrogen technology as a 'black box'.

The stakeholder visioning workshop employed a variety of brainstorming and small group techniques to envision a range of hydrogen energy systems. Building on the outputs from the workshop, insights from the literature review and initial interviews with stakeholders, the authors developed a set of visions which sought to capture the range of prevalent views about what a hydrogen future might or should look like. These visions comprised:

• Structured narrative storylines describing archetypal configurations of hydrogen production, infrastructure (storage and distribution) and end-use technologies

- Indicative quantitative indicators to provide a sense of the scale of technological deployment implied
- Systems diagrams providing pictorial representations of each vision

The credibility, transparency and internal consistency of the visions were then tested through consultation with our stakeholders, and the visions refined. This consultation was also designed to ensure that the visions covered a broad enough range of possible hydrogen futures and that no relevant future was excluded from the subsequent analysis.

It is important to emphasise that the six visions in this study are *not* predictions. Indeed, the technological building blocks for the UKSHEC visions could also be configured into a large number other possible systems, some of which may prove to be much more likely. Instead, the visions are intended to cover the broad range of possibilities in a manageable number of visions. The aim is to promote thinking about the sort of systems that are desirable and achievable in the long term, and to open up discussion around how different hydrogen systems might meet sustainability objectives. This means that the results cannot be seen as advocating or endorsing any one of the visions alone, but they are rather to be thought of as tools for learning about the important technologies, issues, and uncertainties that surround the hydrogen debate.

Multi-criteria mapping

Multi-criteria mapping (MCM) is a multi-criteria appraisal method developed by Stirling with an emphasis on capturing alternative framings and value-based perspectives (Stirling and Mayer, 1999; Stirling, 1999). The approach is based on the understanding that there is not necessarily a single 'best' solution, or as Stirling puts it: 'the aim is to explore the way in which different pictures of strategic choices may change, depending on the view that is taken – not to prescribe a particular "best choice"' (Stirling, 2005b: 5). Multi-criteria mapping thus maps the sensitivities of performance according to different perspectives, uncertainties and framing assumptions.

In terms of process, MCM is conducted through a series of two to three hour, one-to-one interviews with expert stakeholders, using a dedicated software package developed at the Science Policy Research Unit, Sussex University (for details of the interview procedure, see the MCM Interview Protocol, Stirling, 2004). Interviews were recorded and transcribed, providing a rich source of information on participants'

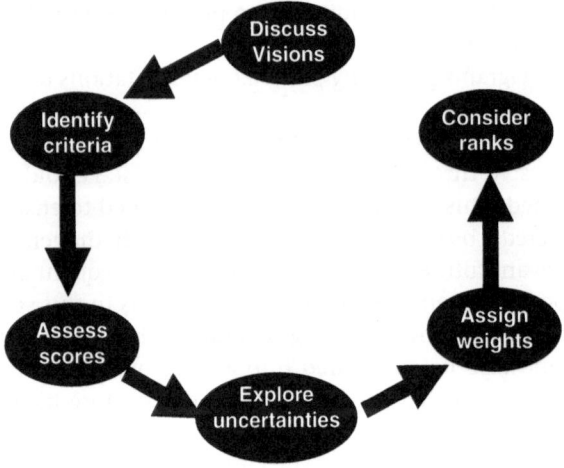

Figure 11.2   The Multi-criteria mapping process

deliberations, reasoning and arguments. A brief overview of the method is outlined below, and further details are available from the MCM Manual (Stirling, 2004).

The interview takes the participant through a structured series of stages, as illustrated in Figure 11.2.

The process allows participants to frame their own appraisal. Participants have the opportunity to define new visions, and define their own criteria for judging the sustainability of the visions. In addition to criteria against which performance can be scored, participants are able to identify issues of principle, against which visions are either acceptable or unacceptable. In the process of scoring, participants are asked to provide scores based on both optimistic and pessimistic assumptions, and to explain the uncertainties that these differences in scores represent. Participants weight the importance of each criterion, and finally, the software produces a visual map of the final weighted scores for each vision.

Visions

Set out below are brief summaries of each of the UKSHEC visions (for descriptions of the full visions see Eames and McDowall (2005).

Central Pipeline

Hydrogen has become the dominant transport fuel, and is produced centrally from a mixture of clean coal and fossil fuels (with

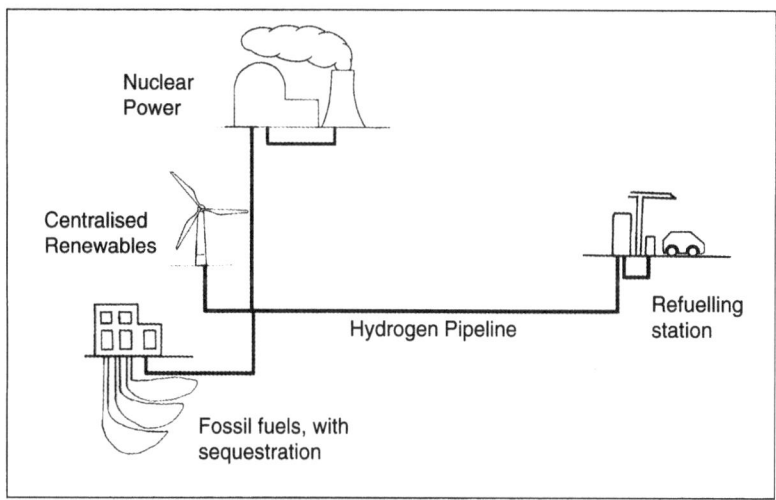

Figure 11.3   Central Pipeline

C-sequestration), nuclear power, and large-scale renewables. Hydrogen is distributed as a gas by dedicated pipeline.

Forecourt Reforming

Hydrogen produced locally from natural gas is the dominant road transport fuel. The existing natural gas network provides the delivery infrastructure, and hydrogen is generated on-site by steam methane reforming at the refuelling station.

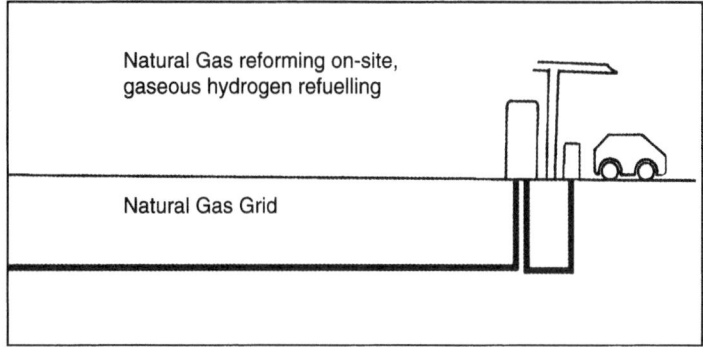

Figure 11.4   Forecourt Reforming

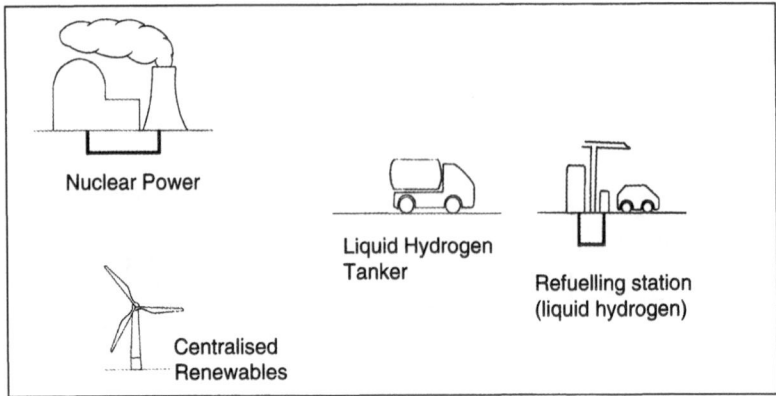

Figure 11.5   Liquid Hydrogen

Liquid Hydrogen

Liquid hydrogen produced by nuclear power and large scale renewable installations has become the dominant transport fuel. There is an international market in liquid hydrogen. This is largely a scenario of substitution, with current energy and transport paradigms remaining unchanged.

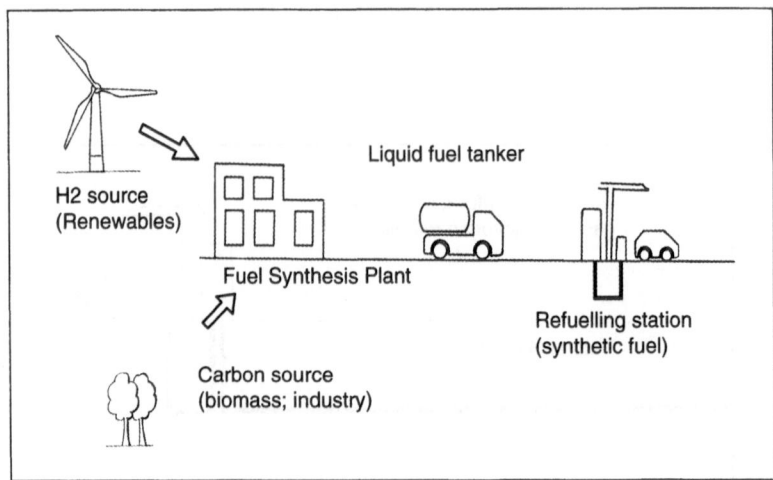

Figure 11.6   Synthetic Liquid Fuel

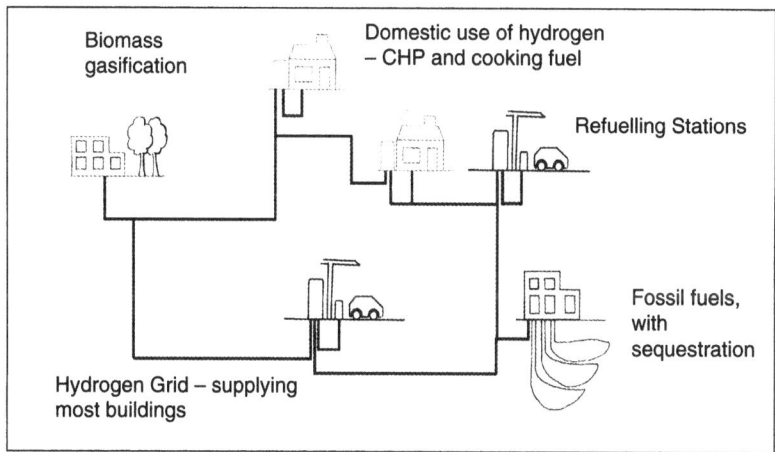

Figure 11.7 Ubiquitous Hydrogen

Synthetic Liquid Fuel

Renewably produced hydrogen again provides the dominant transport fuel. In this case, however, it is 'packaged' in the form of a synthetic liquid hydrocarbon, such as methanol, to overcome the difficulties of hydrogen storage and distribution. The carbon for fuel synthesis comes from biomass and from the flue gases of carbon-intensive industries.

Ubiquitous Hydrogen

Hydrogen, produced through onsite electrolysis, is the dominant road transport fuel, and also plays a vital role overcoming the intermittency problems of a renewables-based electricity system. Hydrogen production is flexible, and can respond to variable electricity supply conditions, easing load-balancing. Since hydrogen is produced onsite it requires no distribution infrastructure. Locally-stored hydrogen provides back-up power for domestic and commercial CHP units at times of peak electricity demand/ limited supply.

Electricity Store

Renewably produced hydrogen is a major energy carrier for heat and power as well as the dominant transport fuel. A hydrogen pipeline grid serves most buildings. Many homes and businesses use fuel cell CHP systems running on hydrogen, and it is common to refuel your vehicle at home. Hydrogen is produced from a mix of larger centralised and smaller-scale distributed renewables and biomass.

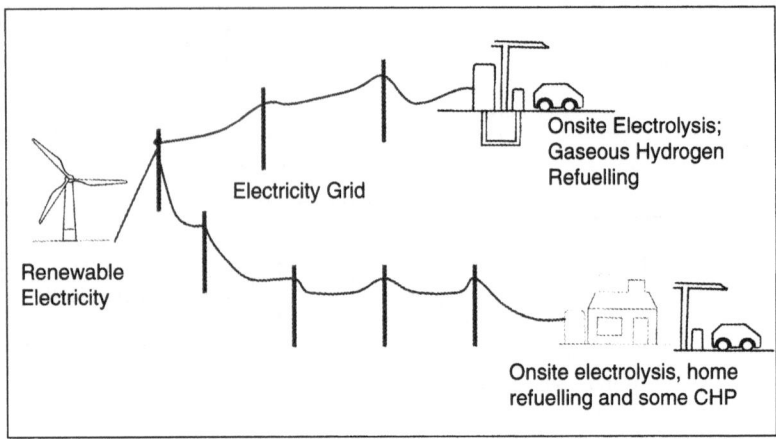

Figure 11.8   Electricity Store

Status Quo

In addition members of the expert panel were also asked to appraise a 'status quo' or reference scenario, describing the current systems UK energy and transport system, which served as a benchmark for comparing the different visions.

Results

The aggregated final weighted scores provide an overall picture of participants' appraisals, showing which visions were thought to be more or less sustainable.

The results from the multi-criteria mapping indicate that, overall, *Electricity Store* was seen as the most sustainable vision, subject to concerns about feasibility. *Forecourt Reforming* was judged to be least sustainable, largely because of carbon emissions, but also concerns about the energy security and economic implications of natural gas dependence.

Central Pipeline was the most contentious vision, with the widest range of rankings, reflecting divergent opinions on nuclear power, carbon sequestration, and the economic viability of a large, centralised pipeline infrastructure. *Synthetic Liquid Fuels* performed unexpectedly well, reflecting the benefits of a low carbon fuel that is straightforward to store and transport, and that offers fewer technological barriers than the use of pure hydrogen. It was also the vision around which there

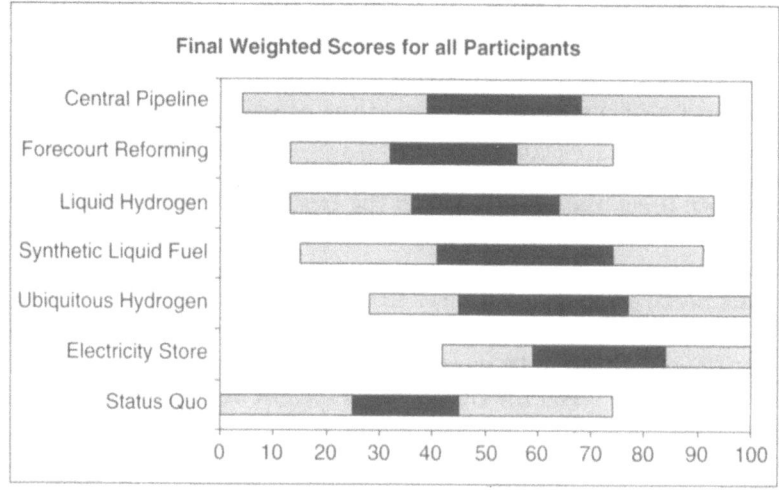

Figure 11.9 Final weighted scores for all participants. Bars indicate extreme (grey) and average (black) pessimistic and optimistic scores, capturing the degree of uncertainty about future performance. The x-axis is a relative scale indicating low (0) to high (100) performance.

was greatest uncertainty. *Liquid Hydrogen* did poorly, partly because of some participants' concerns about nuclear power, but more importantly because liquid hydrogen was seen as impractical and inefficient for use as a mainstream transport fuel (although many participants felt that liquefied hydrogen would have a role in some applications). Finally, *Ubiquitous Hydrogen* performed relatively well, but as with *Electricity Store*, there were some concerns about its feasibility.

As noted above, in addition to the six hydrogen visions, participants were also asked to appraise a *Status Quo* or business as usual vision. It was notable that for many of the participants, there were conditions under which the *Status Quo* was not the worst performing option, implying that some hydrogen futures could be less sustainable than current or business as usual activities. However, *Status Quo* was frequently the worst performing option. In no case was status quo seen as the best performing option, suggesting broad agreement that many hydrogen systems bring sustainability gains.

Of course, the final picture of ranks across all participants tells us little about the different values, uncertainties and framing assumptions that participants brought to the appraisal. These insights are the more meaningful outputs from work attempting to open up the debate around the future sustainability of different hydrogen systems, and it is to these that we now turn.

Criteria and weighting: dimensions of sustainability for hydrogen

Between them, the 15 members of the expert panel defined a total of 98 sustainability criteria, of which many were very similar across different participants (for example, various criteria exploring carbon emissions, social acceptability, energy security, and so on).

The weightings chart provides an overview of the groups of issues that participants judged to be most important. Participants were invited to identify criteria under the classic three elements of sustainability: environmental, economic, and social, as well as an energy policy category of energy security, and an 'other' category for criteria that participants felt did not fall within the other groups. There is a clear tendency for environmental issues to receive high weightings, with social issues in general receiving much less attention, and with a substantial spread of views around the importance of economic criteria.

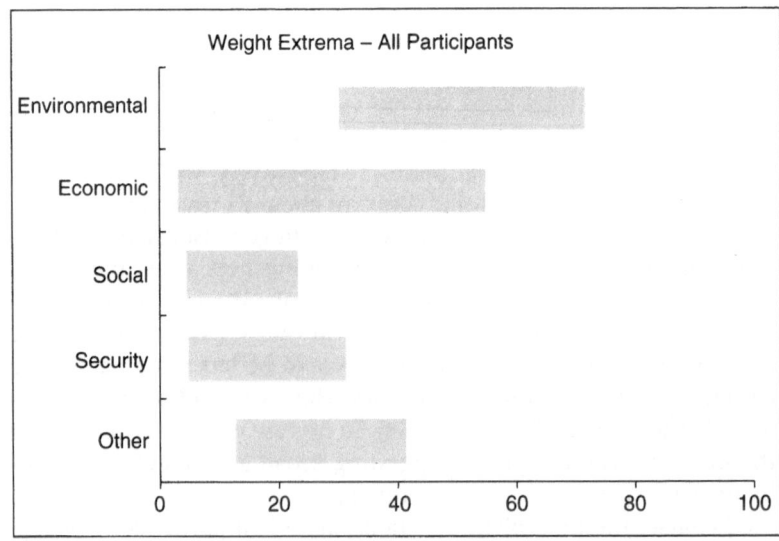

Figure 11.10 Weight extrema for all participants each participant distributed 100 weighting 'points' among their criteria, to indicate relative criteria importance.

Environmental issues

The six visions were clearly differentiated on the basis of their environmental performance. This was dominated by carbon emissions, but included a range of other criteria. In terms of weightings across the participants overall, carbon emissions were clearly considered to be the most important single determinant of sustainability. Environmental issues other than carbon emissions favoured *Electricity Store* and *Forecourt Reforming*.

Issues considered under environmental criteria included:

- Greenhouse gas emissions
- Local air quality
- Toxicity and non-carbon pollution
- Visual impact
- Nuclear waste
- Impacts on water
- Impacts on wilderness
- Impacts on biodiversity
- Catastrophic risk
- Resource depletion

Economic issues

The economic scores are interesting, with none of the visions coming out as obviously better or worse when the appraisals of all participants are examined at an aggregated level, although many individual participants did see significant variation among the visions in terms of economic performance. All participants scored some form of economic criterion.

The most highly weighted economic criteria concerned feasibility, and the economic attractiveness of the vision to investors. Nine participants scored some kind of 'cost' criterion. However, these were varied. Some of these criteria concerned costs to society overall, while others were intended to represent what consumers might pay at the pump. Variations in the assessments of likely economic performance of the visions were in part dependent on different assumptions about policy frameworks around carbon; fossil fuel prices; the costs of nuclear power; and the relative affordability of more decentralised, modular systems or capital-intensive centralised systems.

Issues considered under economic criteria included:

- Cost or affordability of hydrogen
- Impact on the UK economy
- Degree of consumer choice
- Business case/economic feasibility
- Upfront capital costs

Social issues

Seven participants scored only a 'social acceptability' criterion under this heading, and the way in which it was scored suggested that participants felt that this was a potential barrier to feasibility, rather than an ongoing dimension of a desirable or sustainable future. Most participants also gave social acceptability relatively low weightings. The performance of the visions varied amongst participants, with some feeling that 'out of sight' centralised systems such as *Central Pipeline* would be most acceptable, and others feeling that publics would be most willing to accept the least polluting visions, such as *Electricity Store*.

Some participants scored visions on a wide range of social and political concerns. These other social issues tended to be given higher weightings than the more homogenous 'acceptability' concerns. In general, visions involving greater decentralisation tended to do well under these criteria. Issues considered under social criteria included:

- Social or public acceptability
- Access to energy services
- Social justice
- Degree of physical intrusion

- Usability
- Control of energy
- Degree of state intervention required

Energy security issues

All but three participants scored criteria under 'energy security'. Unsurprisingly, *Forecourt Reforming* did badly under energy security criteria, given its dependence on natural gas. Issues considered under energy security criteria included:

- Security of primary sources
- Diversity of primary sources
- Resources scarcity

- Infrastructure and downstream supply
- Compatibility with decentralised systems

Other issues included

- Quality of supply
- Technical feasibility
- Public safety
- Flexibility

- Radioactive Waste (seen as both environmental and social/political)
- Complementarity with renewables
- Geo-political concerns

Uncertainties affecting vision performance

The appraisal as a whole clearly demonstrates the huge uncertainties involved. In some participants' views, the scale of uncertainties within the visions is as important as the differences between them, a conclusion that should not be surprising given the long time horizons involved. The task of the analysis is to explore the basis of those uncertainties.

Uncertainty about technologies

There are uncertainties surrounding technologies, not only in terms of their physical performance, but in terms of what impacts the technologies might have in broader socio-economic terms. The following uncertainties were each identified by more than three participants, and were reflected in variations between pessimistic and optimistic scores:

- Potential leakages of CO_2 from carbon capture and storage
- Fuel cell performance
- Performance of small scale natural gas reformers – in terms of both cost and pollution
- Likely carbon balance and toxic emissions from synthetic liquid fuel synthesis and use
- Costs for all technologies were subject to uncertainty, but in particular uncertainties relating to the costs of synthetic liquid fuels, nuclear power, and pipeline infrastructures were raised
- Significant uncertainties around public acceptability of technologies in general
- Performance, integrity and vulnerability of pipelines
- Very large uncertainties around the possible impacts on the UK economy as a whole

Other areas of uncertainty, raised by fewer participants, included: hydrogen storage, safety of handling hydrogen in a domestic environment, safety of liquid hydrogen, likely developments of fast-breeder reactors (seen as necessary if uranium resource constraints are to be avoided), efficiency of liquefaction, performance of electrolysers, likely pollution from biomass gasification, necessary purity levels of hydrogen for PEM fuel cells, whether the natural gas network can be upgraded to take hydrogen, and whether decentralisation constrains or enhances access to energy.

Sensitivity of vision performance to different possible future contexts

Variation between optimistic and pessimistic scores also occurs where there is uncertainty about the broader context in which the visions exist, such as:

- Future natural gas availability and price – particularly important for the feasibility of *Ubiquitous Hydrogen*, and the feasibility and costs of *Forecourt Reforming*.
- Future national and international climate change policy frameworks, such as carbon taxes, clearly have an important effect on the feasibility of the visions, and on their relative costs.
- Social attitudes towards technology and the environment.

Perspectives and issues in appraising the sustainability of hydrogen

Overall, carbon emissions was clearly felt to be the most important factor on which to judge the sustainability of the different visions. However, participants differed strongly over three key issues, and it is participants' attitudes towards nuclear power, decentralisation, and feasibility that most clearly define different perspectives on how to judge the future sustainability of hydrogen systems. One participant noted:

Nuclear power

> *Nuclear ... is fundamentally opposed to the notion of sustainable development. The idea that you have to bury waste in a hole for a hundred years before you can even deal with it, to me flies in the face of the leaving the world in the state that you found it. However, I see it as a lesser of evils debate, because leaving the world closer to the risk of catastrophic climate change is probably a worse thing to do.*

Some participants saw nuclear power as a necessary and desirable part of a future hydrogen mix. Many others saw nuclear as a 'necessary evil' – a technology that is problematic, but worth the potential difficulties given the challenges of climate change and energy security. Three participants were strongly opposed to nuclear power, one of them ruling out any vision that included nuclear on principle. Their reasons for opposition went beyond concerns about environmental impacts to encompass more social and political concerns. One reason for opposition was a belief that nuclear power is 'anti-democratic', and likely to lead to militarisation of the state. In the context of a future world that will be to some extent destabilised by climate change, expansion of a potentially dangerous technology was viewed as fundamentally undesirable. A second reason was a belief that the development of new nuclear power would in practice mean that renewables and energy efficiency would not be pursued. There were also debates about its cost.

Decentralised systems

I think that hydrogen has the potential to revolutionise ... the way we use energy, by enabling us to produce and manage the supply and distribution and use of energy locally...we can emphasise local control of environment so to some extent we can empower local people in their control over energy services.

There are claims in the hydrogen futures literature and popular press about the potential for hydrogen to enable decentralisation and consumer awareness of energy or even greater democratisation and empowerment. The members of the expert panel took a range of views about such claims, and their approach to decentralisation was an important factor distinguishing their appraisals. Some felt that more decentralised systems would encourage renewables, energy efficiency, and changing consumer behaviour. Others disagreed, and did not see this as likely or plausible, feeling that the way in which the technological system was organised did not imply decentralised structures of ownership, management and control. There were also disagreements about the relative costs of centralised and decentralised systems.

Feasibility, practicality, and speed.

What's important is how quickly will this particular route get to the end game [of low carbon emissions]. And I would say that's probably THE most important issue. Because we might not have very long.

Some participants felt that the most important issue was not to compare the likely sustainability impacts of the various hydrogen systems, since with the partial exception of *Forecourt Reforming*, all the visions tackle the basic problem of climate change. This represents a very different understanding of sustainability from those with strong concerns about nuclear power, for example. The question, for these participants, was more to do with the feasibility and practicality of arriving at the visions. The most important dimension of feasibility was in terms of the economic case for investing in the technology, or the presence of a consumer logic. This raised important questions about the degree to which radical change in the face of environmental pressures is possible in a democratic consumer society.

> *Why should the customer want to do this rather than maintain the status quo? In a democratic situation the customer is going to have to want to do one of these rather than be told to do one of these.*

Mapping different perspectives in the appraisal

The way in which participants approached these three key issues had a major impact on their overall ranking of the visions. Three participants (the *Energy Policy Researcher, Environmental Campaigner,* and *Regional Government Policy Maker*) were strongly opposed to nuclear power, and strongly favoured renewables and decentralised systems. Their appraisals and weightings are shown below.

A second group of participants took a view much more clearly defined by economic feasibility. This group comprised the *Sustainable Energy Policy Consultant, Industrial Gases Industry Participant, Department for Transport Policy Maker, Health and Safety Regulator, Automotive Industry Participant* and *Nuclear Industry Expert.* Some of these participants felt that there would be little difference between the environmental performance of the six visions, with the exception of *Forecourt Reforming.* Instead, the important aspect of appraisal was the relative feasibility and economic attractiveness of the visions.

Striking differences between the patterns of appraisal are clear, based on very different perceptions of what is important in determining sustainability.

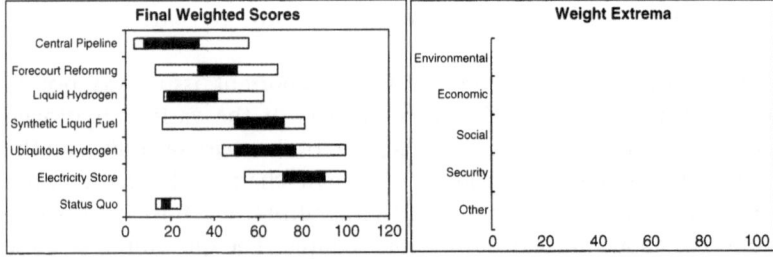

Figure 11.11   Weighted scores and weightings for the *Energy Policy Researcher, Environmental Campaigner* and *Regional Government Policy Maker*

Figure 11.12 Weighted scores and rankings for other participants

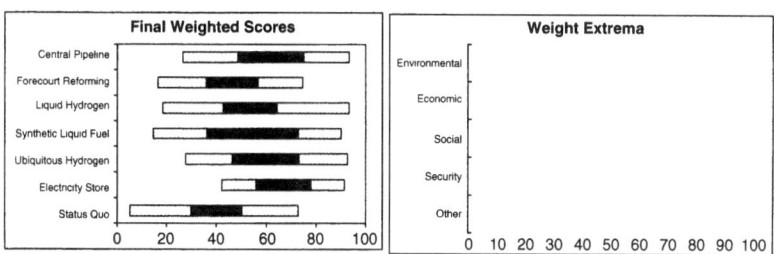

(*Sustainable Energy Policy Consultant, Industrial Gases Industry Participant, Department for Transport Policy Maker, Health and Safety Regulator, Automotive Industry Participant* and *Nuclear Industry Expert*)

Hydrogen appraisal in the UK

In developing its strategic framework for hydrogen energy, the Department for Trade and Industry (DTI) commissioned a consortium of energy policy and engineering consultants to assess different possible hydrogen energy systems for the UK (E4Tech, *et al.*, 2004). This represented a very different sort of process to the one explored here, and comparisons between the two help to illustrate the relative merits of the different approaches.

The consortium appointed by the DTI, led by E4Tech, initiated their work with a series of interviews with senior policy-makers, in an effort to understand the major objectives of energy policy, and thus the criteria on which any hydrogen energy systems were to be judged. E4Tech and colleagues then modelled the cost and carbon impacts of various configurations of hydrogen production, distribution and end use technologies compared with incumbent and alternative systems. Sensitivity analyses explored the impacts of long term uncertainties around oil prices (a base case of $25/barrel and a high oil price case of $50/barrel). Judgements were also made about the impacts of the pathways in terms of the UK's upstream energy security. Six hydrogen energy chains were identified as providing cost effective carbon reductions by 2030. All focus on hydrogen use in transport, distributed by pipeline from centralised hydrogen produced from wind, biomass, coal gasification, reforming of natural gas, nuclear electricity, or possible future 'novel hydrogen production routes'.

The analysis of E4tech *et al.* (2004), provides well grounded, unambiguous advice for policy-makers on the basis of a set of clear assumptions. In doing so, it provides a valuable source of evidence and justification for policy-makers' decisions. Indeed, the UK government accepted many of the report's recommendations. It is an archetypal, and high quality, example of analysis designed to 'close-down' the scope of decision-making, by excluding the less optimal policy options on the basis of a defined set of criteria.

In contrast, the analysis outlined in this chapter seeks to 'open up' the appraisal of hydrogen futures to a broader range of perspectives. The six hydrogen energy pathways identified as the most promising by the E4Tech *et al.* (2004) study would all be variations on the *Central Pipeline* vision outlined in this case study. *Central Pipeline* received the widest array of final weighted scores in this study, performing very well in the views of some participants, but very poorly in others. Where it performs badly, it is on the basis of uncertainties and considerations that were excluded from the E4Tech *et al.* (2004) analysis, such as the social and political sustainability of large, centralised energy infrastructures, or a lack of trust in the ability of certain technologies to perform as promised.

The relative merits of the different approaches are clear. The 'opening up' approach illustrated here provides a broad picture of the pattern of performance of potential futures, against a range of different perspectives. It provides ambiguous advice, and attempts to provide the decision-maker with greater clarity about the potential performance of options given different future conditions and priorities, rather than identify the single best answer. The 'closing down' approach of E4Tech provides clear, evidence-based recommendations, but excludes relevant perspectives and uncertainties, and thus only provides a partial picture of the relative pros and cons of different options. This may explain why the UK Government continues to fund stationary applications of hydrogen in its flagship 'Hydrogen, Fuel Cells and Carbon Abatement Technologies Demonstration Programme', even though the E4Tech analysis explicitly rejected stationary applications as unlikely to help meet government objectives by 2030 except in some niche applications. Similarly, the expert panel that appraised the UKSHEC hydrogen futures in this exercise were all aware of the E4Tech study and its conclusions, but *Electricity Store* and *Ubiquitous Hydrogen* still performed well. Justifications and evidence that lay outside the scope of the E4Tech analysis continue to be important in decision-making.

The two approaches need to be seen as complementary, rather than in opposition. Traditional engineering and economic analysis provides an important, and rigorous, analysis of the performance of options under the dominant assumptions of social priorities, and on the basis of the best data and the best assumptions of experts. Participatory scenario/appraisal methods provide a more general picture of the relative performance of options under a broader range of uncertainties, and a broader range of social views. As such, they provide a systematic and formalised way to integrate social priorities and engineering and economic realities, while remaining sensitive to the inevitable sources of uncertainty.

Conclusions

Debates around energy policy are growing ever more heated, as the impacts of war in the Middle East and economic growth in Asia are felt at the petrol pump, and fears over global warming and peak oil grow more acute. The range of technological solutions proposed to solve these problems is enormous, but none are certain, uncontested, or unarguably sustainable. The situation in which energy policy-makers find themselves today is one in which, in the words of Funtowicz and Ravetz (1993: 744) 'the facts are uncertain, values in dispute, stakes high, and decisions urgent'.

Hybrid methodologies combining participatory foresight with multi-criteria appraisal offer a set of tools to help policy-makers arrive at more robust, democratic, and accountable decisions, based on the best available evidence. This is not to suggest that these approaches make decisions any easier – energy policy will always be controversial, and there will always be conflicts between different, legitimate perspectives that cannot be resolved. Hybrid approaches of the sort illustrated in this chapter recognise this, and seek to provide policy-makers with a clearer picture of the evidence and the perspectives on the debate.

Acknowledgements

The research for the case study described in this chapter was carried out at the Policy Studies Institute, as part of the UK Sustainable Hydrogen Energy Consortium, with funding from the UK Engineering and Physical Sciences Research Council. Special thanks are due to Andy Stirling, at the University of Sussex, for advice and help with the MCM method that he developed, and to Toby Champion at Sussex for help with the MCM analysis software. Thanks are also due to Jenny Yip, for sketching the Vision diagrams.

Notes

1. 'Opening up' is used here in the sense of Stirling (2005a) who makes a distinction between appraisal processes that open up appraisal to a broader range of inputs, and those that 'close down' appraisal by limiting inputs to those than are amenable to quantitative analytic scrutiny.

2. For a more detailed account of the work see McDowall and Eames (2006b) and the series of working papers available from www.psi.org.uk/ukshec

References

K. Anderson, S. Bows, S. Mander, S. Shackley, P. Agnolucci and P. Ekins, *Decarbonising Modern Societies: integrated scenarios process and workshops.* Tyndall Centre for Climate Change Research Technical Report 48: 2006.

F. Berkhout, A. Smith and A. Stirling, 'Socio-technical regimes and transition contexts', in Elzen, B., Geels, F.W. and Green, K. (eds) *System innovation and the transition to sustainability: Theory, evidence and policy* (Camberley: Edward Elgar, 2004).

N. Brown, B. Rappert and A. Webster, *Contested Futures: a sociology of prospective techno-science* (Aldershot: Ashgate, 2000).

DTI, *Energy Projections for the UK: Energy Paper 68* (London: Department for Trade and Industry, 2000).

M. Eames and W. McDowall, UKSHEC Hydrogen Visions. UKSHEC Social Science Working Paper No. 10 (London: Policy Studies Institute, 2005).

EC DG Research, *World Energy, Technology, and Climate Policy Outlook.* Report number 20366 (Brussels: European Commission, 2003).

E4Tech, Element Energy and Eoin Lees, *A strategic framework for hydrogen energy in the UK* (London: A Report to the Department for Trade and Industry, 2004).

Foresight Programme, *Energy for Tomorrow – Powering the 21st Century* (London: Department for Trade and Industry, 2001).

S.O. Funtowicz and J.R. Ravetz, 'Science for the post-normal age', *Futures*, 25, 7 (1993) 739–755.

IEA, *World Energy Outlook 2004* (Paris: OECD, 2004).

R. Kemp, J. Schot and R. Hoogma, 'Regime shifts to sustainability through processes of niche formation: The approach of strategic niche management', *Technology Analysis & Strategic Management*, 10, 2 (1998) 175–195.

D. King, 'Climate Change Science: adapt, mitigate, or ignore?' *Science*, 303 (5655) (2004) 176–177.

L. Kowalski, L. Bohunovsky, R. Madlener, I. Omann and S. Stagl, *Participatory MCE of renewable energy scenarios for Austria – a multi-scale analysis.* Paper presented 17 June 2005, European Society for Ecological Economics, Lisbon, (2005).

W. McDowall and M. Eames, 'Forecasts, scenarios, visions, backcasts and roadmaps to the hydrogen economy: a review of the hydrogen futures literature', *Energy Policy*, 34 (2006a) 1236–1250.

W. McDowall and M. Eames, *Towards a Sustainable Hydrogen Economy: A Multi Criteria Mapping of the UKSHEC Hydrogen Futures – Full Report.* UKSHEC Social Science Working Paper No. 18 (London: Policy Studies Institute, 2006b).

N. Nakicenovic and R. Swart (eds) *Special Report on Emissions Scenarios. Report to the IPCC* (Cambridge: Cambridge University Press, 2000).

J. Robinson, 'Futures under glass: a recipe for people who hate to predict', *Futures*, 22, 8 (1990) 820–842.

Shell International, *Energy Needs, Choices and Possibilities: Scenarios to 2050.* Global Business Environment (London: Shell International, 2001).

A. Stirling, 'The appraisal of sustainability: some problems and possible responses', *Local Environment*, 4, 2 (1999) 111–135.

A. Stirling, *Detailed multi-criteria mapping interview protocol*. Mimeo (Brighton: SPRU, University of Sussex, 2004).

A. Stirling, 'Opening up or closing down: analysis, participation and power in the social appraisal of technology', in Leach, M., Scoones, I. and Wynne, B. *Science, Citizenship and Globalisation* (London: Zed Books, 2005a).

A. Stirling, *Multi-criteria mapping: a detailed analysis manual, Version 2.0*. Mimeo (Brighton: SPRU, University of Sussex, 2005b).

A. Stirling and S. Mayer, *Rethinking Risk: a pilot multi-criteria mapping of a genetically modified crop in agricultural systems in the UK* (Brighton: SPRU/Genewatch, 1999).

H. Van Lente, *Promising Technology: the dynamics of expectations in technological development* (Enschede, Department of Philosophy of Science & Technology, University of Twente, Netherlands, 1993).

J. Watson, A. Tetteh, G. Dutton, A. Bristow, M. Kelly and M. Page. *Hydrogen Futures to 2050*. Tyndall Working Paper 46. Tyndall Centre for Climate Change Research, 2004. http://www.tyndall.ac.uk/publications/tech_reports/tech_reports.shtml [last accessed 3/01/07]

12
Conclusions

Paul Bellaby

This book is about how innovation in technology may or may not gain public acceptability, and whether the process by which it is introduced affects that outcome. The cases that have been considered here are not past but contemporary, even prospective: mobile phones, genetically modified food, nanotechnology and hydrogen energy. In every case there has been controversy, but not of the same kind and by no means to the same degree, which is what makes the comparison instructive.

Public acceptability

'Public acceptability' may mean one of two things: 'in the public interest' and 'acceptable to the public'. They may or may not coincide in the particular case.

From government's standpoint, it is in the public interest to enable innovation to take place, so long it offers economic opportunities or improves quality of life for the greater number. But by its very nature innovation has uncertain consequences. Thus, in implementing new technology government should avoid exposing anyone to unacceptable risks and also avoid excluding sections of the population from the benefits simply because they cannot afford the costs.

From a neo-liberal perspective, however, the market would resolve the issue if left to do so by government, for the market would ensure that each end user or individual consumer would decide in his or her own interest and thereby achieve the greater good.

An un-resolved issue that emerges from each of the chapters in this book is *who* or *what* should define the public interest? First, should it be politics or the market? If it is to be the market, how can the ability of large corporations to monopolise or block innovation in their field be counteracted? And can we trust the market to deliver long-term solutions to such vital problems as global warming and climate change? If politics has to take a hand, should experts, civil servants, politicians, commerce, other stakeholders or citizens be involved? Is it to be some or all of them? If all, are they to have equal or unequal weight?

As Irwin points out, throughout Europe for a considerable period it is ministers and senior civil servants acting on expert advice that have defined the public interest, but this top-down approach has been challenged in the last decade or so. To what extent is the new emphasis on 'public dialogue' about new technologies changing the old decision process? As Irwin argues, dialogue does not guarantee trust, and apparently greater openness does not signify complete recognition of uncertainty.

The range of empirical evidence and discussion presented in this book provides pointers as to how these questions may be answered. It also addresses in some detail an issue that is central to the public acceptability of new technology: the nature of the 'risk' each case may carry, and how the various parties may perceive those risks. Finally it has much to say about the part played by 'trust' (and mistrust) between the same parties.

Can the public be trusted?

The issue of trust is usually cast as how we can encourage the public to trust experts and their political representatives to take decisions for them. Even if power were more equally shared than is the case, trust of authority would be by no means irrelevant, for differences of knowledge are built into specialised divisions of labour, and representative democracy may be the only practical way of ensuring public engagement with many items on the political agenda.

However, as Horlick-Jones points out in this book, experts and policy-makers tend to *mistrust* the public, whatever the public may think of them, believing that the public tends to misunderstand new technologies and to amplify the risks. If experts and policy-makers were correct in this belief, it would be rational to avoid public dialogue and prevent sensitive information from being released.

Opponents call this belief the 'deficit model' of lay knowledge. They argue against it that *experts* typically have a partial view of the consequences of new technology. They 'frame' their attention to include only elements that are significant to the science of the day, and only problems that can be solved by accepted methods. The public, on the other hand, considers a wider range of consequences that new technology may have. This, by itself, justifies greater transparency and public involvement in deciding on the development and implementation of new technologies. In other words, in the eyes of those who oppose the deficit model, the public *can* be trusted.

Knowing risk

In spite of this critique of the deficit model, there can be both differences in forms of knowledge of the same object and deficits in knowledge in each form on the part of various types of actor. This is familiar to those who study expert and lay knowledge of ill health. In this context, it is customary to distinguish between 'disease' – learned knowledge of the psycho-biological process that appears to produce ill-health in the organism, and 'illness' – experiential knowledge of the consequences of ill-health (real or imagined) for one's self. People with chronic conditions typically become 'expert' in handling the consequences day to day, even if they do not have the medical expertise necessary to understand the disease process that produces the effects they experience. In addition, some writers distinguish from disease and illness, 'sickness' – the representation of ill health by the society and culture of which the sick person is a part (for instance, Frankenberg 1980; Turner 1995; Bellaby 1990a, 1999). Representation is integral to communication: for example, there are socially-accepted ways of expressing the experience of illness, that vary with cultural norms, including in some contexts 'suffering in silence' what is in other contexts indulged by the care of others.

Analogously in the energy and other science and engineering fields discussed in this book:

1. Abstract knowledge belongs with science, engineering and technology and its risk assessments.
2. Practical know-how rests with stakeholders – producers, distributors, consumers and regulators – that is, the users of technology.
3. Representation sits with public opinion, opinion leaders and the media.

Table 12.1 implies both differences in forms of knowledge and also deficits on all sides – wherever that form does not belong to the type of actor.

Table 12.1   Difference and deficit in knowledge: type of actor and form of knowledge

Types of agent		Different objects of knowledge/ Varying degrees of knowledge of each		
		Abstract knowledge	Practical experience	Representation
Expert	*Science*	Assessed risks	Deficit in experiential knowledge	Deficit in cultural knowledge
Practitioner	*Stakeholder (consumer, producer)*	Some deficit in science	Preferences and perceived risks	Some deficit in cultural knowledge
Lay	*Opinion Former/ Public*	Deficit in science	Deficit in experiential knowledge	Culturally constructed risks

Knowledge of the physical world is at some level necessary for all who act, but tends to be limited to what one can get by with. For practical purposes no one needs to understand the thermodynamics and mechanics of the internal combustion engine in order to start a car and drive it. The engineer who designs the drive line does need to know these things, but at the same time abstracts from the field, often discarding practical considerations, not arbitrarily, but in order to make the study manageable within the prevailing paradigm.

You have to be a 'stakeholder' (e.g. producer or consumer) for practical experience to be relevant to your actions. Everyone is likely to be a stakeholder in transport in some circumstances, but the car manufacturer is a different case from a scientist and engineer and also from those who have neither abstract knowledge nor a close practical interest in the field. Stakeholders can be expected to have an economic interest in the new technology: that is, a concern for the balance of its benefits and costs and the extent to which it carries risk and uncertainty. Potential producers and distributors of the new technology will look to it as an investment opportunity and seek to assess the market for the product, the competition they face from other producers and distributors, whether government policy may help secure its future. The greater the risk and uncertainty, the higher the rate of return they will seek.

Finally, publics and opinion formers are – comparatively speaking – 'bystanders' with respect to the other forms of knowledge, yet typically construct sense of both the science and the stakeholder's interest in the new technology. Representation at a remove may stigmatise a technology, as it has in the case of nuclear power in the last 20 years or so in the USA and UK. Flynn J. *et al.* (2001) highlight the exceptional stigma associated with nuclear technologies in the USA; Kasperson *et al.* (in Slovic, 2000) demonstrate the impact of such stigma and its media representations in, and effects on, the social amplification of risk. Representation typically associates a technology with such value-laden, affective qualities as 'clean/dirty' or 'safe/unsafe'.

Different approaches to risk analysis are associated with each of these forms of knowledge. Thus, risk assessment conventionally belongs with abstract knowledge. The psychology of decision-making belongs with actors' strategies, preferences and perceptions of risk, which is represented here by several contributors (especially Fischer and Frewer, and Barnett and Timotijevic). Cultural theory takes 'representation' in society and culture as its starting point (Douglas, 1992; Douglas and Wildavsky, 1983; Bellaby, 1990b). Sherry-Brennan and colleagues, for example, show the localised importance of social representations of new technology. Table 12.1 implies the relevance of all three approaches and that they are not necessarily opposed and may be complementary.

There may be a struggle between advocacy coalitions (Sabatier and Jenkins-Smith, 1999) that vie with each other to define how the 'risk' in question should be represented. Hogenboom *et al.* (in Cohen, 2000: 91) emphasise that risk definition is a complex social process involving many different groups: 'risk-producing institutions, government agencies, scientists and environmental organizations. Each of these ... has particular interests and tries to influence risk definition and control in accordance with its unique perspective'.

Finally, Table 12.1 suggests that difference and deficit in knowledge among the various actors in such an arena as consultation about the safety of a new technology are likely to be in a dynamic relation to each other. For instance, a public's trust in the risk assessment offered by a scientist or engineer may be impaired if it comes to light that the risk assessment was funded by a commercial stakeholder, for the representation of the risk in question will be likely to change. Conversely, stakeholders may change their views if it becomes apparent from survey or focus group research, which recasts the representation of the risks in question, that pursuit of a favoured course is unlikely to achieve acceptance by publics.

Risk signatures

Horlick-Jones argues here that risks have 'signatures' that distinguish them, and two faces; risks are 'material' and so constrain what humans can do, and also 'constructs', in that actors confer meaning on them which enables a range of responses, from aversion through to acceptance of risk for the benefit or opportunity that may accompany taking it. Freely adapting this concept to cover benefit, cost and uncertainty as well as aspects of what is more commonly thought of as 'risk', we might classify each of the four new technologies that have been discussed in this book, as suggested in Table 12.2.

Imminence

The first two rows of the Table refer to properties of what is commonly understood by 'risk'. All four cases involve imminent risk, in the sense that users encounter them at close quarters: mobile phones are carried on the person and used in close proximity to the brain; genetically modified foods would be regularly ingested; nano-scale items might also be ingested, whether intentionally or not; and hydrogen energy would accompany us in travel or in our homes and workplaces. By contrast, climate change must seem a remote risk.

Table 12.2 'Risk signatures' of the new technologies covered in the book

| Dimension of risk | New technologies | | | |
	Mobiles	GM food	Nano-technology	H2 energy
Imminence	High	High	High	Moderate
Familiarity	Already present and widespread	Already present but not yet in UK	Prospective only	Prospective but a substitute
Benefits	Adds value for all	Only commerce benefits	Unknown: neutral	Adds value for all
Costs	Known and acceptable	Cheapness irrelevant	Unknown	High now, but later?
Uncertainty	Low	Perceived high	High	Moderate
In sum: is the risk acceptable?	Uncertain but value makes up	Why take the risk?	Unknown but dreaded	Depends on regulators

Familiarity

Familiarity usually inures users to the risks that technologies present. For example, drivers of cars with petrol tanks are at far from negligible risk of explosion in the event of collision or fire in the event of leakage when refilling at a petrol station, but rarely give the matter a thought. The new technologies discussed in this book differ sharply from each other in this respect. Mobile phones are relatively new but by now familiar; GM food is established already in the USA, but has yet to reach Europe and so is not familiar to our target population; nanotechnology is anything but familiar; hydrogen as energy is not familiar, except in thinly distributed demonstration projects, such as the CUTE buses, but it could fit into a familiar niche in both public and private transport as a substitute for petroleum fuels.

Benefits

The next two rows of the table refer to the benefits and costs of the innovation. Again the case studies differ with respect to who (at least in the eyes of the public) is likely to benefit most. Mobile phones are already benefiting millions of personal users, and are usually considered to add value by enabling more flexible communication than both landline phones and PC-based email. GM food had at the time of the notorious consultation been unable to establish a reputation in use in Europe and, worse, was promoted by and so considered to be for the benefit of commercial interests that were not even based in Europe. Nanotechnology is unfamiliar, as we have seen, but also in development within science, and may enjoy thereby a 'neutral' image: of potential benefit to all and not only to a particular interest group that aims to exploit it. Finally, hydrogen energy is represented as a substitute for familiar means of propulsion already in use and – in the course of the last year or two – public interest in the benefits that hydrogen and other low emission renewable fuels promise for stemming climate change has increased quite sharply.

Costs

Costs of any new technology are likely to be a factor in the willingness of consumers to use them, but are widely recognised to be subject to reductions as demand increases. This presumably reinforces the appeal of mobile phones, especially when used for text messages. GM foods and nanotechnology differ from mobile phones in this respect, not, however, because of current or likely future costs, but because cost considerations have so far been outweighed by the risk they are thought to carry. O'Garra *et al.* in this book pursue the issue of cost of hydrogen energy in the context for which this low-emission renewable fuel seems designed: willingness to 'make sacrifices' for the public good – how much more would the public be willing to pay to travel on a hydrogen-powered bus? The issue of cost would be constructed differently if hydrogen were the fuel in regular use, and indeed the material cost would probably be lower as well.

Uncertainty

The fifth row of Table 12.2 addresses the question of how much degree of uncertainty may influence judgements. When a technology is 'new', there is a large degree of uncertainty that is all but absent when a technology has become embedded and regulation has developed after long experience in use. Although, technically, risks are always probable not certain, a high degree of uncertainty is likely to increase anxiety about risk among both stakeholders and publics.

In sum: is the risk acceptable?

The final row in the Table represents the summary judgement: is the risk acceptable? At the time that the research that Mohr reports was conducted there was controversy about frequent use of mobile phones damaging the brain. The controversy seems to have had little if any impact on the spreading use of the technology, presumably because of the other factors discussed above. At that time, and indeed subsequently, the hazards attributed to GM food – both to its consumers and to biodiversity in the eco-system – were uncertain. But the sharply differing signature of GM foods from mobile phones on other dimensions seems to have led the public to the view that no risk was worth taking in this case. A similar conclusion can be reached for nanotechnology: this is not because a commercial interest is seen as a unique beneficiary, but perhaps because the technology is unimaginably small, might be ingested, and so akin to a virus and a source of dread. Hydrogen energy is often thought by its proponents to carry the burden of the misunderstood Hindenburg disaster and the even more misunderstood hydrogen bomb. However, those consulted in the Ricci *et al.* study seem confident that regulators would sort out any safety issues by the time hydrogen came onto the mass market.

Risk signatures and commercial stakeholders

Eames and McDowell in this book look more specifically than others at how stakeholders might approach similar issues. For example, how might someone in business respond to a new technology in its field, given the concept of 'risk' that is relevant to its standpoint and also its orientation to the benefits and costs associated with the new technology? It might have an eye to the competition, and the possibility that a competitor might get in early to exploit what might be an opportunity if it does not act first. It might also have an eye to signals from government and quasi-governmental agencies, like regulators: whether they appear to back the new development, or are uncertain of public reaction and so prefer a consultation first. The competition the business has in mind may be international rather than national as may the reaction of government. Moreover, whether or not the business takes up the new technology may depend upon the scale of its operation as a whole and the percentage of resources the new opportunity would absorb. If bigger fish come to bite, that can absorb any risk, smaller ones may seek out alternative opportunities where they can develop a niche market rather than venture into competition for a mass market. Should government appear to be giving a green light to a new technology, the business will have to assess how trustworthy that lead is and how likely the policy is to be sustained over the period of development of the new technology and its launch in the market.

Consumers, conceived as individuals or households, have some similarities to and some differences from businesses. To be sure, benefits, costs and risks are all relevant to consumers, but their purchases of new technology are as likely to be conditioned by aesthetic preferences or their duty of care to children and partners, as by their individual economic interests. They might also be in competition with others in their social circle and keen to keep a step ahead. Their ability to invest in new technology with this in mind will be conditioned by their income and credit worthiness.

A citizen who has in prospect a new piece of technology in her backyard (neighbourhood), might be especially concerned about risk to health and safety, as the Hornchurch hydrogen filling station suggests which Hodson *et al.* discuss. Businesses plainly not only consider risks, they consider costs, including opportunity costs, and they have an eye to benefits. To a greater or lesser extent, the same applies to consumers. Residents in a neighbourhood affected by new technology, may not only be concerned with risk to their own health and safety but also, as consumers, with whether local property values might be affected by how that risk is represented to potential buyers.

Consultations about new technologies

Governments who seek to introduce new technologies probably mount public consultations with a view to heading off backlash to the innovations. On the other hand, consultation with commercial stakeholders is likely to be aimed at encouraging innovation or the application of new technology. Both forms of consultation are communication processes in which the characteristics of the new technology – including its risks, benefit and costs – are represented and debated. Risk assessments by scientists and engineers and cost-benefit analyses by economists are likely to be drawn on, especially by those who are presenting the case and so have this knowledge at their command, but the same data may be open to opposing interpretations, for instance, to a disagreement about whether a new technology is 'completely safe' or constitutes a 'significant risk'.

Not to consult would assume that lay people would come to accept change as beneficial in due course, even if new technology were forced upon them. It is hard to imagine circumstances in which government would not want to enlist the support of commercial stakeholders in developing innovations, for otherwise the costs and the uncertainties of new development would fall upon government alone.

Not consulting lay people about the introduction of new technologies would be likely to have unintended consequences. What

Table 12.3   Consequences of consultation for publics' acceptance of new technologies

Risk signatures	Form of consultation		
	Upstream	Downstream	None
Imminent	+	(–)	–
Unfamiliar	+	(–)	–
Benefits:			
Few	–	–	–
Everyone	+	+	0
Costs:			
High	(–)	–	–
Low	+	+	0
Uncertainty	–	+	–

Legend: + positive effect; – negative effect; (–) might be negative; 0 neutral these might be presumably depends on the risk signatures of each. Table 12.3 speculates about the consequences for public acceptance of new technologies carrying different risk signatures, should various actions be taken: consultation occurs and is either upstream or downstream, or sensitive information is withheld. This is an over-simplification, but the purpose is to suggest an approach that might be elaborated with further research.

The risk is likely to be acceptable and it will not make a difference if government decides not to consult the public, should the risk be remote and familiar, the innovation benefit everyone, the costs be low and uncertainty absent. Indeed, to consult in such a case might raise suspicion that the risk was unacceptable.

Conversely, should the risk be imminent and unfamiliar, the benefits appear to be limited to a few, the costs high and the uncertainty great, government might be unwise to withold sensitive information rather than consult with the public. But at what point: upstream (prior to significant development of the innovation) or downstream (when it is ready to be implemented)? The only circumstances in which leaving consultation until the development is almost a *fait accompli* would be less likely to lead to rejection for the project is when uncertainty is high. Consulting early in such a case might jeopardise development that progressively reduces the uncertainty.

Being consulted seems likely to make a positive difference for public acceptance of innovation, but only when the risk signature requires it. Suppose the risk signature does indicate the need for consultation, then how should that be done? There is no one best method. Each has its disadvantages, but also each has the edge in certain applications.

Possible methods range from a poll, whether by census or probability sample, with such minimal involvement as casting a vote, through answering a written questionnaire and responding to a brief oral interview with standardised questions, all the way up to participating in a qualitative investigation, using purposive sampling so as to accentuate variation in type of response, and semi-structured one-to-one interviews or focus groups.

Where little is known by investigators about public response, and by publics about the topic, qualitative investigation is superior. The balance of advantage shifts to polling for consultation after the public has become familiar with the topic. Then a representative view is key.

Of course, those who respond to enquiries of both types tend to trust the democratic process in general, and that this particular process is genuine, not just spin. If others who do not respond prove critical for implementation of a new technology, there is little that either method can offer.

A final word

Engaging with both publics and stakeholders has become *de rigueur* in developing new technologies. However, one important implication of many of the chapters in this book is that consulting with publics and even stakeholders about innovations in technology is not likely to make easy gains for those who are promoting them. As Ricci *et al.* point out it is more likely to encourage debate and questioning than to guarantee acceptance. Furthermore, when taken together, the cases dealt with in the book suggest that whether consultation proves positive at all and whether it is better go upstream than downstream depend on the risk signature of the case. Both the point of consultation (if any) and how any consultation is carried out, is itself a technology that is as yet imperfectly understood. But this book suggests ways in which that technology at least might be developed.

References

P. Bellaby, 'What is a genuine sickness? The relation between work-discipline and the sick role in a pottery factory', *Sociology of Health and Illness*, 12.1 (1990a): 47–68.

P. Bellaby, 'To risk or not to risk? Uses and limitations of Mary Douglas on risk-acceptability for understanding health and safety at work and road accidents', *Sociological Review*, 38.3 (1990b): 465–483.

P. Bellaby, *Sick from Work: the body in employment* (London: Ashgate, 1999).

M. Douglas and A. Wildavsky, *Risk and Culture* (Berkeley, Ca: University of California Press, 1983).

M. Douglas (ed.) *Risk and Blame: essays in cultural theory* (London: Routledge, 1992).

J. Flynn, P. Slovic and H. Kunreuther, Preface in Flynn, J. Slovic, P. and Kunreuther, H. (eds) *Risk, Media and Stigma* (London: Earthscan Publications, 2001).

R.J. Frankenberg, 'Medical anthropology and development – a theoretical analysis', *Social Science and Medicine*, 14 (4B) (1980): 197–207.

J. Hogenboom, A. Mol and G. Spaargaren, 'Dealing with environmental risks in reflexive modernity', in Cohen, M. (ed.) *Risk in the Modern Age* (Basingstoke: Macmillan, 2000).

R. Kasperson, O. Renn, P. Slovic, H. Brown, J. Emel, R. Goble, J. Kasperson and S. Ratcik, 'The social amplification of risk: a conceptual framework', in P. Slovic *et al.* (eds) *The Perception of Risk* (London: Earthscan Publications, 2000).

P. Sabatier and H. Jenkins-Smith (1999), 'The advocacy coalition framework', In Sabatier, P.A. (ed.) *Theories of the Policy Process* (Boulder, Colorado: Westview Press, 1999) pp. 207–221.

B.S. Turner with C. Sampson, *Medical Power and Social Knowledge* (2nd Edition, London: Sage, 1995).

Index